Housing and Home Management

Housing and Home Management

Sangeeta Rani

RANDOM PUBLICATIONS
NEW DELHI (INDIA)

Housing and Home Management

ISBN 978-93-5111-844-2

Published in 2016 in India by

RANDOM PUBLICATIONS

4376-A/4B, Gali Murari Lal, Ansari Road
New Delhi-110 002
Phone : +9111-43580356, 011-23289044, 011-43142548
e-mail: sales@randompublications.com,
info@randompublications.com, randomexports@gmail.com

Reprinted 2026

Type Setting by : Friends Media, Delhi-110089

Preface

Housing generally refers to the social problem of ensuring that members of society have a home in which to live, whether this is a house, or some other kind of dwelling, lodging, or shelter. Many governments have a department that deals with housing, such as the United States Department of Housing and Urban Development. Housing Development Board in Singapore and Housing Authority in Hong Kong. The term informal housing can include any form of shelter or settlement which is illegal, falls outside of government control or regulation, or is not afforded protection by the state.

As such, the informal housing industry is part of the informal sector. To have informal housing status is to exist in 'a state of deregulation, one where the ownership, use, and purpose of land cannot be fixed and mapped according to any prescribed set of regulations or the law.' While there is no global unified law of property ownership typically, the informal occupant or community will lack security of tenure and, with this, ready or reliable access to civic amenities. Due to the informal nature of occupancy, the state will typically be unable to extract rent or land taxes.

The term informal housing is useful in capturing informal populations other than those living slum settlements or shanty towns, which are defined more narrowly by the UN Habitat as 'contiguous settlement where the inhabitants are characterizes as having inadequate housing and basic services...often not recognised or addressed by the public authorities an integral or equal part of the city.' Common categories or terms for informal housing include slums, slum settlements, shanty towns, squats, homelessness and pavement dwellers.

– Author

Contents

1

Introduction

AN OVERVIEW

The operation of a dining room and kitchen presents the best opportunity for the make a profit and, at the same time, presents the best possibility for losing large sums of money. Generally speaking, large amounts of money are involved in the kitchen operation. More important, however, is the food business.

It is of particular importance to your health and well-being of your Brothers who eat in the dining room are naturally affected by the food served them. A kitchen operation should not be undertaken with less than thirty people who will be eating and paying 'board' costs. Because of fixed expenses that are independent of the number of people eating, operating a kitchen with less than thirty 'men' is financially not feasible.

KITCHEN OPERATION

Every topic that is operating a kitchen should either elect or appoint a kitchen steward. The proper supervision of the kitchen is sufficiently demanding in time that it justifies having a member responsible for its management.

The steward should sign all bills and/or receipts when turning them over to the Treasurer for payment or record keeping purposes. The Treasurer then has the authority to make the necessary disbursements.

The responsibility of the steward is strategic when considering that he is the one person who will either make the income for the kitchen 'cover' expenses or have the kitchen realize a deficit. The Treasurer must formulate and discuss the kitchen budget with the steward. It must be determined exactly what amount may be spent for food on a weekly basis.

The Treasurer must, from time to time, check up on the status of kitchen costs with the steward to make sure he is remaining within his allotted budget. Only by using measures designed to impose strict controls on the operation of the kitchen can the unit be kept from incurring a large deficit.

MENUS

The most frequent source of complaint in houses is the food. The reason for this is usually not in its preparation, but rather in the choice of menus. A good steward will plan well balanced, varied menus, and will garnish his main meals with relishes, appetizers, soups and salads; all of which are inexpensive and have often proved to be excellent investments in good will and nutrition. Avoid a set menu for each day of the week. Nothing hurts morale more than "Thursday is macaroni and cheese day."

HOW TO LOSE MONEY IN THE KITCHEN?

There are several well established ways in which a kitchen operation can lose money. The first is by the patronage of dishonest vendors who will bill you for a certain weight or quantity of merchandise, and then won't deliver the correct amount. The steward should arrange delivery times with the vendors so that he may personally inspect each delivery. If the delivery is insufficient according to the delivery slip, the slip should be changed before the steward signs it, acknowledging receipt of the merchandise. When buying in less than full case quantities, a count should be made carefully at the time of delivery to ascertain that the amount or quantity is as stated on the delivery receipt or invoice.

A second area of loss is when the kitchen 'help' frequently takes home considerable quantities of food when they leave the house at the end of the day. It is taken for granted that kitchen 'help' eat their meals while on the premises. It is, however, known that some people who work in kitchens take home enough food to supply their entire families. This can constitute a considerable drain on finances. To prevent a third area of loss, arrangements should be made to keep the kitchen and food storage area locked at all times when the cook is not on duty. It is a sad fact, but true, that disappearance of food and supplies costs every unit of money out of their kitchen budget. A fourth loss is the practice of permitting members to eat at odd hours or between meals. Definite serving hours must be established for each meal and the kitchen declared "off limits" at all other times.

GUEST MEALS

Each topic should define and regulate guest meals in their bylaws. Such meals are equally the evening or Sunday meal, which normally constitute 60 percent of the daily meal cost per "boarder", and any charge for such meals should be based on that percentage, as a minimum. Any member who abuses a guest privilege is costing each 'boarder'; thus, he is being unfair to the its members. There are cases where a member not living in the house wishes to eat certain meals on a regular basis at the house. Such rates may be established by using the daily meal cost per boarder, and using 60 percent of that amount to be charged for evening or Sunday meals, 25 percent of that amount for the noon meal and 15 percent of that amount for breakfast. Even then, a member is not paying his full share and it is reasonable to add 10 percent to these rates.

The practice of refunding for missed meals can become very involved and should not be done unless there is an unusual situation or it is for an extended period of time. Since most Treasurers have had little or no training in food management, consideration of the following six topics would be helpful in obtaining a more efficient and less costly dining facility in your kitchen.

- *Labor*:
  - Hire reliable cooks, follow up on job application references.
  - Pay liveable wages.
  - Work out acceptable working hours (no more than 48 hours a week).
  - Laws are very strict with regard to payment of overtime. Know the laws!
  - Don't make unusual demands.
  - Have some incentive plans.
  - Give pay raises, when appropriate.
  - Provide good physical working conditions.
  - Listen to employee's problems—don't ignore their suggestions.
  - Place yourself in their position.
  - Set ground rules and be sure both sides understand them.
  - Let employees know about vacations, wages, days off, etc.
  - Let employees know to whom they are to report (chain of command).
  - Compliment them when they have done a good job.
  - Cost of labor should never exceed 30 percent of your total board income.
- *Purchasing*:
  - Don't buy from everyone—be selective.
  - Know what you are getting.
  - Designate one person to do the buying.
  - Get price lists from vendors.
  - It is not always necessary to buy Premium, Prime, Grade A, etc., in food items.
  - Don't buy more than needed.
  - Take advantage of good buys.
  - Know delivery times.
  - Don't accept poor merchandise.
  - Try to set up food purchasing on competitive bids—shop around.
  - Food cost should not exceed 50 percent of your board income.
  - Utilize or create a food purchasing cooperative.

- *Equipment*:
  - Take care of equipment.
  - Do 'preventable' maintenance when possible.
  - Replace worn out equipment as soon as possible.
  - Follow company instructions on equipment as to cleaning, safety, etc.
  - Watch for warranty on equipment.
  - Purchase labor saving devices.
  - Purchase commercial grade equipment, not home grade equipment.
- *Sanitation*:
  - Don't ever let this aspect of the kitchen operation deteriorate.
  - Be strict in this area (you'll never be sorry).
  - Give the cook assistance in this area.
  - If bothered by rats, mice or insects, call in a qualified exterminator.
  - Discard or replace equipment that is unsanitary.
  - Have plenty of hot water.
- *Charges and Collections:*
  - Collect from all who are not paying for 'board' but eating occasional meals.
  - For extra functions, the kitchen account should be reimbursed by the account sponsoring the event (i.e., social, rush, homecoming).
  - Don't put anyone on a free meal basis.
- *Administration*:
  - Take a monthly inventory of kitchen supplies and food.
  - Are you having excessive leftovers?
  - Strictly enforce rules on notifying cook when people won't be eating a particular meal.
  - Your total board cost per man should not exceed that charged by the dormitories.
  - Set up kitchen policy similar to dormitories.
  - Consider increasing board prices following increases in food and prices or cut down on meals, quantity or service.
  - Have your cook sign a year long contract.
- *Caution*: In many college towns, unscrupulous fraternity cooks, to whom the inefficient steward leaves the buying, receive personal rebates,either in money or merchandise. Too much emphasis cannot be put on the importance of checking up regularly on the kitchen ?help?, of switching dealers occasionally and of checking regularly on deliveries as they come in.

EMPLOYING KITCHEN EMPLOYEES

The employment of a knowledgeable, reliable and likeable cook should not be a hasty, poorly planned process. Your topic is entering into a strictly business relationship and you should approach the hiring of your cook in that manner.

The following are tips when hiring kitchen employees:

- Have the applicant fill out a simple job application. The job application serves as a source of information, both during the interview and following employment (if that should happen).
- Check out job and personal references listed on the application. Don't take anything for granted. Find out what kind of employee your applicant was, job attendance record and general attitude.
- The applicant should be personally interviewed by the President and Treasurer. If they are in agreement that the applicant is qualified, the advisor should personally interview the applicant.
- If all are in agreement that hiring the applicant is in the best interest, employ the applicant.
- Before employing an applicant, make sure there is agreement on wages, time off (including vacations), insurance, working hours, job expectations and sick days.
- In attempting to establish a fair wage, check with the other fraternities and sororities on campus, and the dormitory food service, to determine what they are paying their employees—that will be the fair wage.
- The dormitory food service is a good source of potential employees. Don't be afraid to advertise in the local newspaper.
- Refer to Internal Revenue Service guidelines in order to comply with federal and state employment laws.

EMPLOYMENT CONTRACT

There are 'pros' and 'cons' to using an employment contract that should be obvious.

'Pros' to utilizing a contract when employing kitchen personnel:

- The employee has a legal responsibility to fulfill commitments stated in the contract.
- The contract specifies job expectations and wages so as to avoid future disputes.
- It lets the employee know you mean 'business'.

'Cons' to utilizing a contract when employing kitchen personnel:

- The topic has a legal responsibility to employ the person in accordance with the conditions of the contract and limits the 'grounds for discharging' an employee.

- Regardless of the status of the kitchen account, the topic must pay the employee in accordance with the conditions of the contract.

FOOD PURCHASING COOPERATIVES

Many campuses and/or Greek Systems operate a food purchasing 'coop'. These coops are mostly referred to as a Fraternity Management Association *(which may offer additional services)* or simply, the Fraternity CoOp, and can save thousands of dollars every year by buying in quantity.

Because of the wide array of these coops, both in operation and quality, it is recommended that you investigate it, if it exists on your campus. If one doesn't exist, have you thought about starting one? The operation of a kitchen facility is no small task. However, if done properly. Additionally, it can produce substantial profits for the properly managed. Most certainly, it provides an opportunity for members to develop their management, decision making and leadership skills.

EIGHT STEPS TO EFFECTIVE PURCHASING OF FOOD AND SUPPLIES

STEP 1: DETERMINE DESIRED IMAGE OF FOOD QUALITY AND SERVICE

An image is the memory or an imagined picture of service, food quality, atmosphere, comparative prices and other characteristics of the operation which you want your customers to have. A poor image reflects poor service, food quality, etc. Ask people what they like. Comments can help determine your image. Do you want to serve high quality food and create that type of image? Or, do you want to serve only low quality food or something between high and low?

STEP 2: KNOW CUSTOMERS' FOOD PREFERENCES

Knowing food preferences specifically is part of Step 1, but deserves special attention. The 'customer' is always 'right' is not the answer, but if you want to improve your menu, customer information is certainly necessary. Would your customers like one or several changes in your operation? How do you know? Asking customers for their suggestions and comments is one way. Try to chat with a customer once in awhile, and after a few discussions, a pattern may develop which can be quite revealing.

STEP 3: ESTABLISH ONE PERSON. FOOD PURCHASING CONTROL AND AUTHORITY WITH RESPONSIBILITY

One person control of purchasing will help maintain consistency of product and profitability through more effective cost control. This means one person can concentrate on food grades and prices, quality specifications, market conditions, vendor evaluations and other necessary purchasing activities. If one person is given the authority to purchase, make sure there is a second in command who can take over in the temporary or permanent absence of the buyer. One person control will promote fast decision making, faster competitive decisions and reduce frustrations normally associated with committee decisions.

STEP 4: MAINTAIN INVENTORY CONTROL

Accurate inventory control is required to determine accurate food cost figures. It should and can be maintained by perpetual records or by physical stocktaking at regular intervals. Inventory records should be maintained for permanent reference in a stock book with space provided for each item. This will help cut losses, control food costs and reduce write offs due to damaged or outdated merchandise. When food costs get out of line, inventory control may indicate pilferage or significant differences between book inventory levels and actual physical product counts.

STEP 5: UNDERSTAND AND KNOW FOOD COST CONTROL PROCEDURES AND FOOD COSTS

A knowledge of food cost is absolutely necessary in order to determine the total cost of operating, total profits, possible kitchen shrink, economical minimum order size and, more importantly, menu price. A recent survey of 60 firms found that 16 of the 60 (27%) did not know what the ingredients of their menu were costing them, and, consequently, had no sound basis on which to price out their menus. Competition was probably setting menu prices. Change in food cost as a percentage of sales is a possible indicator of shrinking and improper kitchen control. Changes in food cost must be studied for menu pricing and portion control.

STEP 6: COMMUNICATE WITH SUPPLIER: KNOW PRODUCTS AND AVAILABILITY IN YOUR AREA

Smart purchasing is done when you know what is available. Smart selling by distributors is accomplished when they know what ingredients and products you need.

STEP 7: CHECK ALL DELIVERIES

A purchasing system includes buying, receiving and checking the merchandise upon delivery. While the kitchen follows up on product quality, another check by a third person is advisable as an anticollusion, antipilferage, control device.

Checking and receiving systems include:

- Trained and competent personnel—more than one person if possible.
- Adequate space for proper checking.

- Equipment designed to provide efficient product movement.
- Specifications to be used as standards of quality.
- Scheduled hours, if possible, in order to insure the availability of a checker when delivery is made.

When merchandise is delivered, it should be placed in a safe, quality control location immediately upon off-loading by the driver. Count all unit sale items and weigh all high value items sold by weight; randomly weigh all other items.

Visually check quality of produce and packaged items where quality is essential for compliance with invoice description and/or specifications. Move all items immediately to storage areas and lock if immediate access is not necessary. Compare invoice prices against purchase order. Verify and mark (stamp or signature) all receiving invoices. Send all documents to buyers for further verification and book inventory update.

STEP 8: EVALUATE AND IMPROVE

Menu prices result in part from a supplier?s ingredient cost and a foodservice operator?s labor cost. This cost can be reduced by increasing the number of units handled per dollar of supplier labor cost and of foodservice operator cost. This increased productivity should lower the per serving cost to the consumer, provided hourly or fixed labor costs can be used in several alternative functions or can be transferred to other productive activities.

In order to reduce the per unit cost of handling food both by the supplier and the foodservice operator, there are several questions which an operator should answer:

- Can order size be increased?
- Can more cases be handled per man hour and per dollar of labor cost?
- Can there be fewer but larger deliveries?
- Can job performance be improved?
- Can costly labor functions be performed by competent but lower wage employees?

Labor Transfer: Research revealed that in 58% of deliveries, the driver stored the merchandise. This means that a supplier would certainly realize a substantial savings if the driver unloaded only and the foodservice personnel stored. The foodservice operator would realize an increase in cost. Due to wage differences between the truck driver and the foodservice personnel, an obvious cost savings would accrue to the delivery system. It was projected that this annual savings would amount to $592.

Decrease Deliveries: Handling efficiency and labor productivity can be improved by reducing the number of deliveries by onehalf, and reducing the number of suppliers by onehalf. This assumes an average weekly number of cases of 283. Or the same number of suppliers could be used, but alternate weekly deliveries be scheduled from each supplier, or at least half the normal weekly deliveries would be scheduled. This would result in a net gain of 8.25 man-hours per 1,000, or an annual savings of $709.

Single Delivery: The onestop concept of delivery means simply that a supplier provides all the necessary products to a foodservice operation. All products for a given week, for example, would be on one invoice, representing one delivery. There is a potential savings of 12.02 man-hours per 1,000 cases, or an annual savings of $1,184. The supplier would realize savings of 9.14 man-hours per 1,000 cases, and an annual dollar savings per establishment of $1,074. The operator would realize savings of 3.49 man-hours per 1,000 cases, and an annual dollar savings of $110.

No additional storage space would be required. The space utilization factor for the 60 foodservice operators surveyed was approximately 6070%. One stop is a feasible alternative.

ESSENTIAL DUTIES/RESPONSIBILITIES

- Gathers all dirty dishware from tubs brought to the dishroom, rinses and stacks in dishracks, always clearing dishes of leftovers and trash in order to prepare for automated washing.
- Loads full dishracks on belt and ensures correct placement in accordance to the instructions for the operation of the machine. Operates automatic dish washing machine by using controls as instructed.
- Loads dishwashing solutions into automatic dispensers and regulates the output of detergents and chemicals to the dishwashing machine by monitoring and adjusting controls.
- Unloads clean dishes from racks after being processed through dish machine, inspects for cleanliness and carefully stacks in specified carts and shelves for the easy use by restaurant and kitchen staff.
- Manually scrubs pots, pans and other kitchen equipment that cannot be washed automatically, using detergents, scourers, and special solutions as required. Inspects for cleanliness, manually dries with hand towels, and puts away in correct places.
- Manually polishes all silver, stainless steel and pewter used in food service, using standard polishing cloth and products, in order to present clean and attractive equipment to diners.
- Maintains the automatic dishwashing equipment in good condition and working order in accordance with manufacturers' instructions in order to prolong the life of the equipment, prevent breakdowns and to meet mandated health compliance regulations.
- Keeps the dishroom in clean and orderly condition at all times by sweeping, mopping, emptying trash, wiping counters and equipment

and organizing shelves in order to maintain a sanitary work station and to meet mandated health compliance requirements.

- Mops kitchen floors as requested and at end of each shift, cleans all kitchen work surfaces as regularly scheduled by manager including walls, ceilings, hoods, vents and ovens.
- Gathers all trash cans from kitchen work areas at end of each shift and empties into outside trash compactor.
- Sweeps loading dock and kitchen entrance areas, clearing litter and debris to trash bins, for the safety of employees and purveyors.
- Performs other duties as assigned, requested or deemed necessary by management.

OTHER DUTIES/RESPONSIBILITIES

- Contacts purveyors and contractors, in writing or by phone, when in need of product or service.
- Unloads deliveries as they arrive, and accounts for ordered items by comparing packing lists to actual items received. Transports deliveries to stock room by carrying boxes or using a dolly. Assists the stock room personnel with sorting and storing of delivered goods.
- Maintains clean employee breakroom and restrooms by sweeping, mopping, cleaning counters and refilling supplies on a daily basis.
- Provides assistance to other employees and departments to contribute to the best overall performance of the department and hotel.

TRAINING

STEWARDING TO THE TOP

Kitchen stewarding is growing to be one of the most sought after jobs, with qualified youngsters taking on the industry with an extremely professional approach and redefining the role of a kitchen steward.

CHANGING TRENDS

Earlier it was the onus of the chef to plan the menu, keep the kitchen clean, manage costs and budget, look into aspects of hygiene and, of course, cook. But with the increasing thrust on different cuisines, the role of chefs became more specific leading to a growing need to hire cleaners to do the job. Supervisors were later required to oversee the work of cleaners. Slowly, they evolved to become stewards, taking care of every aspect of the kitchen from maintenance, cleaning, tracking inventory, keeping an eye on the temperature in which utensils, pots and crockery are washed, refrigeration of food and monitoring the work of the cleaners. Today, a kitchen steward complements the role of a chef and the F&B manager. He helps the chef purchase the necessary ingredients and stocks the food the right way and at the right temperature, while helping the F&B manager control the inventory and checking breakages. The kitchen steward has lent professionalism, dignity and technicality to what was previously considered a menial chore.

Prakash Vishwanathan, kitchen stewarding executive, Courtyard Marriott "A few years ago, we didn't need sound knowledge to be in the industry. Today, it is not just about doing dishes and mopping the floors; there are host of technical things one needs to be informed about to handle the job - from choosing the right kind of kitchen equipment including the cooking equipment to cleaning the dishes, and the pots. We also require a certain amount of accounting information to help buy cutlery and hire labour." A steward is the only person in touch with the entire operations of a kitchen. Maintenance and cleanliness aside, he also understands the know-how of food presentation while serving the guests.

THE CAREER PATH

There are no specific qualifications required to be a kitchen steward. Hotels like Le Royal Meridien insist that it is the attitude they look for while recruiting management trainees in their hotel. Explains Paul Dhas, HR, director, Le Meridien, "The trainee should have the right attitude towards the kitchen, the hotel and even the guests, as we also let them handle guests. The first year, we focus on kitchen training, then cross training in the second year, where they handle the responsibilities of a chef and the kitchen. The final year, they don the chef's hat."

Different hotels have different hierarchical policies. After two years of experience, a kitchen steward could go on to become a supervisor and then a chief steward, who is in charge of all the kitchens in the hotel. There are hotels which dispense elaborate hierarchies and have only supervisors and executive stewards. Because of their exposure to cooking, some stewards also go on to become executive chefs of hotels. A working knowledge of the kitchen and cooking also gives stewards an edge to become the restaurant manager, as they are also involved in planning of the menu. The more enterprising ones make a successful career as a general manager.

PAY SCALES

A kitchen steward starts on a salary of Rs 3,000 and a kitchen supervisor earns about Rs 7,000, while an executive steward can draw anywhere between Rs 10,000 and Rs 12,000 per month. A chief steward is in similar rank as an F&B manager and is indispensable to a hotel. His remuneration is anywhere between Rs 30,000 and Rs 40,000 per month. So, if you are thinking of specialising in kitchen stewarding, do not dismiss the option as something monotonous and mundane. The opportunities are plenty and the horizons can be explored. There are success stories in the industry to testify this fact. Go ahead and give it your best shot.

Some of the broad roles of a kitchen steward:

- Cleaning and organising dishes including buffet
- Controling inventory.
- Monitoring the use of machines for different kinds of utensils like pots and crockery.
- Maintaining hygiene and quality in the kitchen and different equipment used there.
- Keeping a tab on breakage and missing cutlery.

FEMALE AND MALE SERVANTS

FEMALE SERVANTS

UPPER SERVANTS

- *Housekeeper*: In households where domestics employed number over twenty-five, the housekeeper's sole duty is to engage, manage and dismiss the female servants, with the exception of lady's maid, nurse and cook, whom the mistress engages. In smaller households, the housekeeper manages the stores, both ordering and dispersing them. She tends to the house linen, both repairing it and replacing it as necessary. She supervises the china-closet, the stillroom department, and superintends the arrangement of bedrooms for visitors and their servants. Her daily routine includes: overlooks the stillroom, sees what china and linen is given out for breakfast, presides over the housekeeper's room breakfast, gives out the stores for the day, assist in washing china, makes rounds of the bedrooms and replaces supplies such as candles, writing paper and soap, makes sure the rooms are clean and in order, presides over the servant's hall dinner, arranges dessert for dinner, makes tea in the afternoon, and makes the coffee for dinner. She also makes preserves and bottles fruit. She keeps the household accounts, and does most of the needlework. In smaller households, the cook often assumes the duties of the housekeeper.
- *Lady's Maid*: A lady's maid attends to her mistress's appearance. She arranges her hair and assists in dressing her. She packs and unpacks the mistress when traveling. She may also make her mistress's dresses. Depending on the size of the household, she may assume some of the housekeeper's duties. In a typical day, she: brings up hot water as necessary, brings up tea before breakfast, prepares clothes for dressing, assists the mistress in dressing, puts the room in order, puts out necessities for walking, riding or driving, assists in

taking off her outdoor attire, puts evening dress in order, assists in dressing her for dinner, sits up for her, assists in undressing her, puts away her jewels, keeps her wardrobe in repair and washes the lace and fine linens. She also attends to any pets the mistress may have.

SENIOR SERVANTS

- *Governess*: A governess taught the children of middle and upper class households until they were old enough to go away to school, college, or to a private tutor. She was generally a well-educated middle-class girl who needed to earn her own living. But although she was expected to have the bearing and education of a 'lady' she was treated as a servant. This often left her in limbo—neither an insider or an outsider, as the other servants resented her as too educated and too good for their ranks.
- *Nurse*: The nurse is in charge of caring for the household's children from the time they are born, until they are turned over to the care of the governess. She washes and dresses the children, feeds them, takes them on outings, and puts them to bed. She makes the children's ordinary under-clothing, and repairs their general clothing. Most nurses have dinner brought to them in the nursery, but some dined with the other servants.

UNDER SERVANTS

- *Cook*: In large households, only the cooking proper is the duty of the cook. All ingredients are prepared for her use by the kitchen maids. A first-class cook attends to the family breakfast after having her own. She makes out the menu for luncheon and dinner, which is sometimes reviewed and altered by the mistress. In town, she orders from the tradespeople who serve the house. She prepares the soup for the following day, prepares the pastry, jellies, creams and entrees for the day, all in the morning. The afternoon is usually her free time, unless there is a dinner party or guests. She then prepares dinner, and once dinner is served, her duties are over for the day. It is also her duty to lock the doors and windows of the basement, to let the kitchen fire burn low, and to turn off the gas in the kitchen and passages before retiring. In smaller households, the cook assumes the duties of the head kitchen-maid and even scullery maid.
- *Kitchen Maid*: In large households, the head kitchen maid is an under-cook and assumes many of the plain-cooking responsibilities. In small households, the kitchen maid prepares vegetables, game and poultry, does the dairy-work, and bakes the bread. If there is no stillroom

maid, she makes the cakes for luncheon, tea and dessert and the rolls for breakfast. She keeps the kitchen clean and keeps things in order.

- *Housemaid*: In large households, the upper housemaid undertook lighter jobs such as making beds and tidying bedrooms. She made sure rooms were supplied with the necessary linens, and that they were kept in repair. She dusted the china ornaments, and tended to the flower arrangements. She kept an eye on the lower housemaids, who would light the fires, clean the living rooms, polish the brass, carry water upstairs for washing, and empty the chamberpots. Some maids were assigned to specific rooms, such as the still-room, laundry, dairy or nursery.
- *Scullery Maid*: Her chief duty is to clean and scour the pots and pans, as well as the cooking utensils. She cleans the scullery, servant's hall, larders, and kitchen passages. She usually dines in the kitchen with the kitchen maid.

MALE SERVANTS

UPPER SERVANTS

- *House Steward*: A House Steward is employed only in larger households where the accounts are too extensive for the Housekeeper to manage. The House Steward has a sitting-room for his duties of household accounting. He may also performance as a Land Steward. Those households having Land Stewards give them their own separate dwelling. The House Steward engages men and women servants, with the exception of the family, ladies' maids, nurses and valet. He pays their wages and dismisses them. He orders household goods, pays the household bills and keeps the household books. He usually submits the household books to his master once a month for review. He does not wear livery.
- *Valet*: Valets are generally kept by single gentlemen and elderly gentlemen. A butler may performance as a valet for a single man. A valet brushes his master's clothes, cleans his boots, carries up the water for his bath, puts out his clothes for dressing, shaves him if necessary, assists him in dressing, packs and unpacks his clothes when traveling. He also loads his rifle when shooting, stands behind his master's chair at dinner, waits at his breakfast and luncheon, attends to the master's wardrobe and sees that everything is in repair and order. A valet to an elderly gentleman attends to his health needs also, and may sleep in the room with his master. He does not wear livery.

- *Butler*: The butler is the head of his department and responsible for the performance of those under him (the footmen). He has usually served his apprenticeship in domestic service, slowly working his way up the hierarchy. His responsibilities increase with the size of his establishment. He is in charge of the plate chest and makes sure it is properly cleaned before use. He keeps accounts of the wine handed out and consumed by the household. He decants the wine for luncheon and dinner, and puts away decanters after each meal. He also bottles wine, and country butlers brew beer. A butler takes over the valet's duty when there isn't one in the household. A butler announces visitors during the afternoon hours. He readies rooms for use every day, as well as tidies them. In households with only one footman, the butler assumes some of the pantry work.

Senior Servants

- *Coachman*: His duties vary depending on the number of footmen employed, and whether or not there is a second-coachman on staff. In families with more than one coachman, the head coachman drives a pair of horses and the second coachman drives one horse. Nightwork is the duty of the second coachman. The head coachman supervises those under him (second coachman and grooms), and sees that the horses are properly fed and taken care of. He also has charge of the the stables and is responsible for ordering supplies. He assists the groom in cleaning the carriages and harness. In some families, coachmen have their meals with the servants. In others, they have their own rooms in the stables.
- *Head Gardener*: The head gardener is in charge of the hot-houses, green-houses and conservatories on the estate. He supervises the rest of the gardeners, their number depending on the size of the gardens.

UNDER SERVANTS

- *Footman*: A typical day for a footman is the following routine: He takes coals to the sitting-room, cleans the boots, trims the lamp wicks, cleans the plate, lays the breakfast table, carries in breakfast, waits at breakfast, removes breakfast, answers the door in the morning after 12 o'clock, delivers notes, lays the luncheon table, takes in and waits luncheon, clears the table and cleans the silver, lays the dinner table, goes out with the carriage in the afternoon, attends to fires throughout the day and evening, prepares table for tea, cleans up after tea, waits at dinner, clears the dinner table, helps clean the plate, washes the glass and silver used at dinner, takes in coffee and dessert after dinner, waits in attendance in front hall when dinner guests are leaving, attends to the gentlemen in the smoking room, attends to lighting in the house at dusk, goes out with the carriage in the evening and valets the young gentlemen in the family. Footmen dress in livery. When one footman is employed, the butler assists in his duties. When two footmen are kept in lieu of a butler and footman, the head footman assumes the duties of the butler. When two or three footmen are kept with a butler, the head footman is called an under-butler, although he remains in livery.

- *Groom*: He attends to the horses and exercises them. He cleans the carriages and harness, and feeds the horses. He also readies the stables for the master's inspection each morning.

ORGANISATION STRUCTURE

Organizing, the process of structuring human and physical resources in order to accomplish organizational objectives, involves dividing tasks into jobs, specifying the appropriate department for each job, determining the optimum number of jobs in each department, and delegating authority within and among departments. One of the most critical challenges facing lodging managers today is the development of a responsive organizational structure that is committed to quality.

The framework of jobs and departments that make up any organization must be directed toward achieving the organization's objectives. In other words, the structure of a lodging business must be consistent with its strategy. Managers give structure to a hotel and lodging through job specialization, organization, and establishment of patterns of authority and span of control.

JOB SPECIALIZATION

There are as many degrees of job specialization within the lodging industry as there are types of organizations, there are many types of organizations. One extreme is the case of a hotel where the owner/operator is responsible for checking in the guests, servicing their needs, taking care of the housekeeping for the guest rooms, maintaining the building and grounds, and checking out the guests. There is, to be sure, much to recommend this method of work. It is rewarding to have total control over a project from beginning to end, and many people find it motivating to the results of their efforts.

However, as the demand for additional products or services increases it becomes more and more difficult for an individual to do his or her job well. One benefit of the increased workload is increased revenue, which would enable the individual hotel operator to add housekeeping staff, one or more front desk agents to check in and check out the additional guests, and engineering and maintenance personnel to care for the building and grounds. As a general rule, specialization increases worker productivity and efficiency. On the other hand, delegating jobs increases the need for managerial control and coordination. Someone has to make sure that housekeeping staff come in after the painters have repainted a room not before! A crucial element of hotel and lodging management is coordinating the many specialized functions within hotels so that the organization runs smoothly.

Specialization has its own set of problems; it can result in workers performing the same tasks over and over again. A point can be reached where the degree of specialization so narrows a job's scope that the worker finds little joy or satisfaction in it. Signs of overspecialization include workers' loss of interest, lowered morale, increasing error rate, and reduction in service and product quality.

One solution to this problem is to modify jobs so that teams can perform them. Instead of a single guest room attendant being assigned to a group of rooms, a work team in a hotel housekeeping department might clean all of the rooms on a particular floor. Some establishments use teams regularly throughout the organization; others use teams more selectively. Teams can be directed by a manager or can be selfmanaged. The idea behind self-managed work teams is for workers to become their own managers, which increases their self-reliance as well as develops a talent pool.

A concept called the quality circle is based on the belief that the people who actually do the work, rather than their managers, are the ones who are best able to identify, analyze, and correct problems they encounter. The idea originated in Japan in 1962. The quality circle is a group of employees, usually fewer than ten, who perform similar jobs and meet once per week to discuss their work, identify problems, and present possible solutions to those problems. For example, a quality circle might be formed among front desk agents. The group forwards its findings and proposals to management for evaluation and action. Quality circles are most successful when they are part of an organization-wide improvement effort. American business picked up on the quality circle concept in the mid-1970s.

THE ORGANIZATION OF A LODGING ESTABLISHMENT

As their facilities grow in size, lodging managers are faced with the need to group certain jobs in order to ensure efficient coordination and control of activities. These job groupings are usually called departments. In general, departments might be grouped as front of the house (those departments in which employees have guest contact, such as front desk), and back of the house (where employees have little guest contact, such as accounting).

However, separating departments by function is the most common method of organizing a hotel or a lodging business. The departmental structures of a limited-service hotel, a full-service hotel with under 500 rooms, and a full-service hotel with over 500 rooms. There may be as few as 2 or as many as 50 employees in a particular department. In a very small lodging business, such as a bed-and-breakfast, the owner can supervise each department. However, as the lodging business increases in size it is most effective to create managerial positions within departments.

ROOM DEPARTMENT

Typically, the rooms department includes reservations, the front office, housekeeping, and telephone or PBX. In smaller full-service hotels, security and engineering might also be included in the rooms department. Responsibilities of the rooms department include reservations, guest reception, room assignment, tracking the status of rooms, prompt forwarding of mail and phone messages, security, housekeeping of guest rooms and public spaces such as lobbies, and answering guests' questions. To perform these many duties effectively, the rooms department may be divided into a number of specialized subunits.

To complicate matters, in many instances these subunits are also referred to as departments. For example, the laundry department is responsible for cleaning and pressing all the hotel's linens and employee uniforms as well as guest laundry. Because of its specialized function, little of the knowledge and skills required to manage a laundry operation is transferable to other areas of hotel operations.

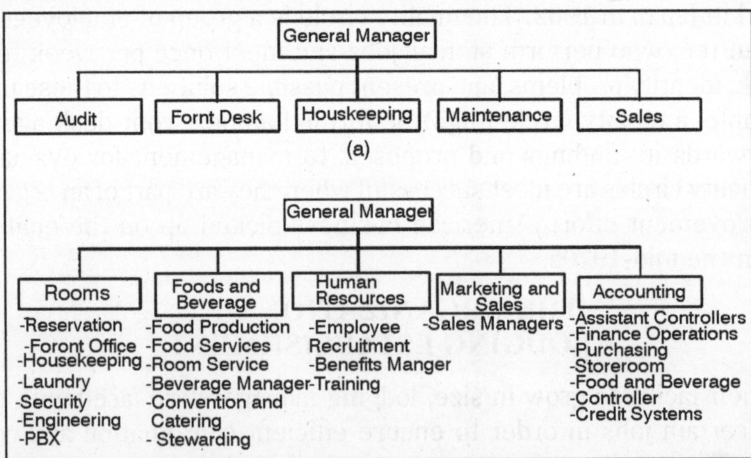

Fig. Department Structure in the Hotel and Lodging Industry: (a) Departments of a Limited-Servoce Hotel; (b) Departments of a Full-Service Hotel (Under 500 Rooms)

The front office is one of the most important departments in a hotel, as it often offers the only contact between guests and staff. A hotel's front office is where guests are greeted when they arrive, where they are registered and assigned to a room, and where they check out. Usually, the telephone operator, other guest communications functions, and the bell staff or those employees responsible for delivering luggage and messages and attending to special guest requests also fall under the front office umbrella. The reservations department takes and tracks the hotel's future bookings. The housekeeping department is responsible for cleaning guest rooms and public spaces. Because of their specialized nature, the security and engineering departments are discussed in separate sections.

A great deal of interdependence exists among the subunits of the rooms department. For example, reservations must inform the front office of the number of presold rooms each day to ensure that a current inventory of salable rooms is always available. On the other hand, the front office must let reservations know whenever walk-in guests register. A similar level of cooperation is required between the front office and housekeeping. When a guest checks out, the front office must inform housekeeping so that the room may be cleaned.

Once a room is cleaned, housekeeping must inform the front office so that the room may be sold. Certain tasks within the rooms department must occur in a specific order. For example, housekeeping cannot properly provision a guest room if the laundry does not supply enough clean towels or bed sheets. Engineering cannot replace a defective light switch in a guest room if housekeeping does not report the problem. Effective management of this busy department calls for standardized plans, procedures, schedules, and deadlines, as well as frequent direct communication between the executives who manage the key operating units of the rooms department.

SECURITY

The hotel and lodging business is vulnerable to security and safety problems. Problems can be created by guests, employees, or intruders. Security breaches can result in embezzlement, theft, arson, robbery, and even terrorism. Depending on the size of a hotel or a lodging establishment, the security function may be handled by a fully staffed department on site, contracted to an outside security company, or assigned to designated staff members or on-premises supervisory personnel in the rooms department. In a larger, full-service hotel, the director of security may report directly to the general manager. In smaller hotels, the security function might become a task of the rooms department.

ENGINEERING

Typically, the engineering department's responsibilities include preventive maintenance; repair; replacement; improvement and modification to furniture, fixtures, and equipment (FFE); and ensuring uninterrupted provision of utilities (gas, electricity, water). Preventive maintenance involves routine checks and inspection of the key components of all equipment. Maintenance of recreational facilities may be part of the engineering department's responsibilities. In particular, swimming pools require extensive maintenance to ensure proper filtration and to prevent the accumulation of algae and other conditions unsuitable for swimming.

Prompt repair minimizes loss of productivity in other hotel operating departments and inconvenience to hotel guests. When a particular FFE has reached the end of its useful life and repair is no longer cost-effective, replacement is indicated. Improvement projects enhance the existing operation or reduce operating costs of the facility. Modification projects alter the existing operation to accommodate one or more new functions.

One hotel might have a large engineering staff that includes plumbers, carpenters, painters, electricians, and other technicians. Another might have maintenance personnel who have general knowledge and understanding of the hotel's operations but rely on outside contractors for specialized jobs. In larger, full-service hotels, engineering may be a separate department, with a director who reports directly to the resident manager.

FOOD AND BEVERAGE DEPARTMENT

The primary function of the food and beverage department is to provide food and drink to a hotel's guests. In earlier times, when an inn had a single dining room that could hold a limited number of guests, this was a fairly simple task. Today, however, providing food and drink is much more complicated. A large hotel might well have a coffee shop, a gourmet restaurant, a poolside snack bar, room service, two banquet halls, and ten function rooms where food and beverages are served. It might also have a lounge, a nightclub, and a lobby bar. On a busy day, it's quite likely that functions will be booked in many outlets at the same time. In addition, some outlets may have multiple events scheduled for a single day. There is great diversity in the types of activities performed by a food and beverage department, requiring a significant variety of skills on the part of its workers.

Because of the diversity of services provided, the food and beverage department is typically split into subunits. The executive chef, a person of considerable importance and authority in any full-service hotel, runs the food production, or kitchen, department. A variety of culinary specialists who are responsible for different aspects of food preparation report to the executive chef.

The actual serving of food in a large hotel's restaurants is usually the responsibility of a separate department, headed by the assistant food and beverage director. The food service department is composed of the individual restaurant and outlet managers, maitre d's, waiters, waitresses, and bus help.

Because of their special duties and concerns, many large hotels have a separate subunit that is responsible only for room service. Because of the high value and profit margins associated with the sale of alcoholic beverages, some hotels have a separate department that assumes responsibility for all outlets where alcoholic beverages are sold.

The person responsible for this department is the beverage manager. Most full-service hotels also do a considerable convention and catering business. The typical convention uses small function rooms for meetings and larger rooms for general sessions, trade shows, and banquets. As a hotel or lodging business increases the use of its facilities for conventions and meetings, it may form a separate convention services department.

The convention services department and its personnel are introduced to the client, a meeting planner, or an association executive by the marketing and sales department. The convention services department then handles all of the client's meeting and catering requirements. Individually catered events include parties, wedding receptions, business meetings, and other functions held by groups. To provide for the unique needs of these types of customers, hotels often organize separate catering and convention departments.

Depending on the size of the hotel, the job of cleaning the food and beverage outlets themselves as well as of washing pots and pans, dishes, glasses, and utensils is often delegated to a subunit known as the stewarding department.

It is only through continuous cooperation and coordination that a hotel's food service function can be carried out effectively. A guest who is dining in a hotel restaurant requires the joint efforts of the kitchen, food service, beverage, and stewarding departments. A convention banquet cannot be held without the efforts of the convention and catering department along with the food production, beverage, and stewarding departments. The sequence of events and cooperation required among the food and beverage staff is even more important than in the rooms department, thus increasing the importance of communication between managers and employees alike. Another challenge faced by management is the diversity of the employees in the food and beverage department; the dishwasher in the stewarding department is at a dramatically different level than the sous chef in the kitchen.

MARKETING AND SALES DEPARTMENT

Coordination is not as important an issue in, the marketing and sales department which is generally much smaller than the food and beverage department. The primary responsibility of the sales managers who make up the marketing and sales department is sales, or the selling of the hotel facilities and services to individuals and groups. Sales managers sell rooms, food, and beverages to potential clients through advertising, attendance at association and conference meetings, and direct contacts. The marketing and sales department is also removed from most of the day-to-day operational problems faced by other departments. The division of work among the sales managers is based on the type of customers a hotel is attempting to attract. Individual sales managers often specialize in corporate accounts, conventions, or tour and travel markets. Sales managers' accounts are sometimes subdivided along geographical lines into regional or national accounts. The sales staff of the largest full-service hotels usually does not exceed a dozen or so. These sales managers work more or less independently in their particular market segments.

HUMAN RESOURCE DEPARTMENT

The human resources department serves no customers, books no business, and prepares no meals, yet it plays a vital role in a hotel's efficient operation. The three functions of the human resources department are employee recruitment, benefits administration, and training. The director of human resources is also expected to be an expert on federal and state labor laws and to advise managers in other departments on these topics. The human resources department's major challenge is in its interactions with other hotel departments.

Although the human resources department recruits, interviews, and screens prospective employees, the final hiring decision rests within the department in which the potential employee will be working. The same is true of promotion and disciplinary decisions; the human resources department's input is, in most cases, limited to advice and interpretation of legal questions. The human resources department's effectiveness depends on its manager's ability to form effective working relationships with managers of other departments.

ACCOUNTING DEPARTMENT

In many hotels, the accounting department combines staff functions and line functions, or those functions directly responsible for servicing guests. The accounting department's traditional role is recording financial transactions, preparing and interpreting financial statements, and providing the managers of other departments with timely reports of operating results. Other responsibilities, carried out by the assistant controller for finance, include payroll preparation, accounts receivable, and accounts payable.

Another dimension of the accounting department's responsibilities deals with various aspects of hotel operations, cost accounting, and cost control throughout the hotel. The two areas of central concern to the accounting department are rooms and food and beverage. The accounting department's front office cashier is responsible for tracking all charges to guest accounts. At the close of each business day, which varies by hotel but typically occurs at midnight or after the bulk of guests' transactions have been completed, the night auditor is responsible for reconciling all guest bills with the charges from the various hotel departments. Although the front office cashier and the night auditor physically work at the front desk and, in the case of the cashier, have direct contact with guests, they are members of the accounting department and report to the assistant controller of operations.

The food and beverage department may be responsible for food preparation and service, but the accounting department is responsible for collecting revenues. The food and beverage controller and the food and beverage cashiers keep track of both the revenues and expenses of the food and beverage department. The food and beverage controller's job is to verify the accuracy and reasonableness of all food and beverage revenues.

In addition to tracking and preparing daily reports on the costs of the food and beverages used in the hotel, in many cases the accounting department is also responsible for purchasing and storeroom operations. Finally, the director of systems is responsible for designing the accounting and control systems used throughout the hotel. The accounting department is anything but a passive staff unit contending with routine recordkeeping. The accounting department is also responsible for collecting and reporting most of a hotel's operational and financial statistics, which provide important data for decision making and budget preparation purposes.

The head of the accounting department may report not only to the hotel's general manager but also to the hotel chain's financial vice president or to the hotel's owner. The reason for this dual responsibility and reporting relationship is to afford the hotel corporation an independent verification of the financial and operating results of the hotel.

GENERAL MANAGER

In addition to being in charge of overseeing all of the departments that we have discussed, the hotel's general manager (GM) is responsible for defining and interpreting the policies established by top management.

The general manager serves as a liaison to the hotel's owner or corporate parent, sets the overall strategic course of the hotel, sets hotel-wide goals, coordinates activities between departments, and arbitrates interdepartmental disputes. It is common practice in a large, full-service hotel for a director of public relations to report directly to the GM. The GM also has corporate-level responsibilities, participates on civic boards and committees, and engages in industry-related activities such as serving on the local tourism commission or hotel-motel association.

In addition to possessing a high level of technical skill (i.e., a thorough understanding of each operating department in the hotel), the general manager must also be decisive, analytical, and skilled with both computers and people. He or she must be able to see the big picture and how all of the parts of the hotel fit into the overall organization.

RESIDENT MANAGER

An executive may be promoted to relieve the general manager of some operational duties. This is often accomplished by elevating the duties and responsibilities of one particular department head without relieving that person of regular departmental duties. The title of this position is usually resident manager. It is quite common (and logical) for the general manager to select the manager of the rooms department to be resident manager. Responsibilities of the resident manager include serving as acting GM in the GM's absence, representing the GM on interdepartmental hotel committees, and taking responsibility for important special projects such as major hotel renovations, VIP guests, and operating reports that require in-depth analysis for the regional or corporate offices.

PATTERNS OF AUTHORITY

The delegation of authority creates a chain of command, the formal channel that defines the lines of authority from the top to the bottom of an organization. The chain of command consists of a series of relationships from the highest position in the organization to the lowest. The chain of command specifies a clear reporting relationship for each person in the organization and should be followed in both downward and upward communication. Following the chain of command enables each new employee, no matter what his or her position, to know exactly for whom and to whom he or she is responsible.

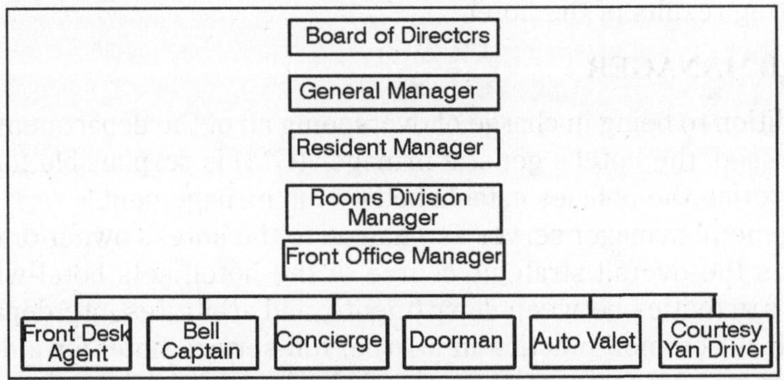

Fig. Chain of Commeand

When designing an organizational structure, managers must consider the distribution of authority. Defined simply, authority is the organizationally sanctioned right to make a decision. Authority can be distributed throughout an organization or held in the hands of a few select employees. Decentralization is the process of distributing authority throughout an organization. In a decentralized organization, an organization member has the right to make a decision without obtaining approval from a higher-level manager. Centralization is the retention of decision-making authority by a high-level manager.

Traditionally, hotel and lodging management has been very centralized, probably due to its roots in small, owner-operated lodging. In recent years, as the hotel and lodging industry has expanded, decentralization has become a more frequent style of operation.

Decentralization has several advantages. Managers are encouraged to develop decision-making skills, which help them advance in their careers. The autonomy afforded by this style of operation also increases job satisfaction and motivation. When employees are encouraged to perform well, the profitability of the organization increases.

Many hotel and lodging organizations have begun to empower employees and supervisors to make decisions that typically have been made by managers. One example that we have already discussed is the use of the quality circle. For example, if a front desk agent determines that a guest's bill is incorrect, in a decentralized organization the agent has the power to make the correction immediately. If that same front desk agent determines that a guest's stay has been unsatisfactory, he or she has the power to reduce the guest's bill by an amount previously specified by management. Additional challenges, control of the process, and quality assessment become part of everyone's job, and each employee is given the authority to take positive actions that will lead to high quality and improved performance.

SPAN OF CONTROL

Span of control refers to the number of people who report to one manager or supervisor. A wide span of control results in a flat organization- that is, a large number of employees reporting to one supervisor. A narrow span of control results in a tall organization, in which a small number of employees report to a supervisor, necessitating a larger number of supervisors. No formula exists for determining the ideal span of control. The following factors determine the most appropriate span of control: task similarity, training and professionalism, task certainty, frequency of interaction, task integration, and physical dispersion.

When a large number of employees perform similar tasks, the span of control can be increased. When the employees perform very different tasks, the supervisor must give each subordinate more individual attention in order to keep in touch with the different types of tasks; this requires a narrower span of control. For example, the rooms department manager might easily manage the front desk agents and housekeepers until the brand standards for check-in or checkout of a guest increase in complexity and the standards for the various types of rooms and their cleaning procedure increase in detail. At this time, the rooms department manager's span of control must be narrowed.

The better trained and more skilled a subordinate is, the less supervision is required. For example, a front desk agent might require a higher level of training and skill than a room service waiter. Thus, a front desk supervisor can supervise more employees (wider span of control) than the room service supervisor (narrower span of control).

Task certainty refers to the predictability of a task. Routine tasks allow management to devise standard procedures for subordinates to follow, minimizing questions about the job and widening the span of control. On the other hand, close supervision is called for when tasks are ambiguous and uncertainty is great. For example, the task of checking a guest in or out of the hotel can be documented and standard procedures can be created, so the front desk manager can have a relatively wide span of control over the front desk agents.

However, because of the diversity of customers the sales manager encounters, the tasks of a hotel sales manager are less certain. The director of sales is responsible for coaching the sales managers, observing sales calls, and ensuring deal closings; these tasks require a hands-on approach that limits the number of employees the director of sales can handle, narrowing the span of control.

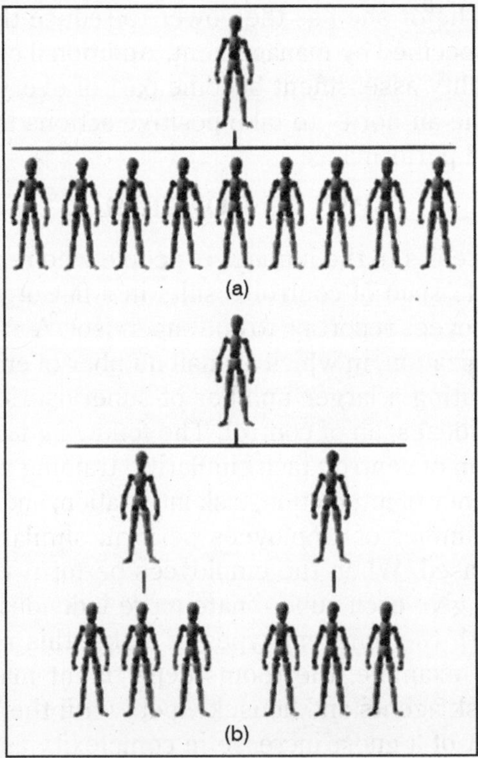

Fig. Span of Control: (a) Wide Span of Control (Flat Organization): (b) Narrow Span of Control (tall Organization)

If the supervisor-subordinate relationship requires frequent interaction, the span of control must be narrow. If interaction is infrequent, the span of control can be wide. For example, the hotel controller must review regularly the status of collections and payments with staff.

In contrast, the frequency of interaction between the chief engineer and his or her subordinates is directed by written work orders in most circumstances and thus requires less direct communication, enabling a wider span of control.

If the supervisor must integrate and coordinate the tasks of subordinates, the span of control narrows. For example, in the production of a meal for one table of four guests with different appetizers, salads, and entrées, the chef must have a narrow span of control to ensure that each component of the meal is assembled correctly and delivered to service personnel on a timely basis. The span of control of the executive housekeeper can be much wider because the procedure for cleaning and preparing each guest room is similar, if not exactly the same, for every room.

Physical dispersion refers to the distribution of employees within the lodging establishment. For example, if the executive housekeeper has guest room attendants on 20 floors in two buildings and the front desk manager has all front desk agents located in one place, the span of control would narrow for the executive housekeeper and widen for the front desk manager.

The ideal number of people that one person can supervise depends on a variety of factors. Consistent with trends in organizational structure such as teams, quality circles, and employee empowerment, many hotel and lodging organizations are widening their span of control. The objective behind these trends is to develop a flatter, more responsive organizational structure in which employees can make decisions without going through several levels of management.

THE PROS AND CONS OF FUNCTIONAL ORGANIZATION DESIGN

The most important strength of a functional organizational design is efficiency. The performance of common tasks allows for work specialization, which increases overall productivity. Workers develop specialized skills and knowledge more rapidly. Training is easier because of the similarity of tasks and the resulting opportunities for inexperienced workers to learn from experienced workers. This helps new employees quickly learn the kinds of behavior that lead to success and promotion. Coordination of activities within functional departments is easier than in more broadly based organizations.

A functional organization fosters efficiency, teamwork, and coordination of activities within individual units. However, the functional organization's most important strength is also the source of its great est shortcoming. The success of a hotel as a business is measured by its overall performance and not by the performance of any one department. A hotel with spotless guest rooms will not be successful if guests' front desk experiences are not up to par. Even if guests' dining experiences are superb, the hotel will fall flat on its face if its rooms are dismal. It is sometimes difficult for each department to fully appreciate its role in the overall success of the organization. It is vital that each department keep in mind the hotel-wide goals of customer service and profitability rather than focus narrowly on its own concerns. Some means must be found to coordinate the activities of functional departments and to set hotel-wide strategies and goals. A hotel's functional organization demands strong leadership.

New initiatives in hotels often require cooperation and coordination between functional departments. New ideas tend to be stillborn if department heads lack a hotel-wide perspective or have difficulty coordinating their activities. The tendency to concentrate on doing things right often overshadows the organization's ability to do the right thing.

The GM, as the chief executive of the hotel, emerges as the single person capable of providing the overall organizational direction, decision making, coordination, and arbitration needed to make a hotel's functional departments work together effectively. A hotel simply cannot run itself from the departmental level. Thus, the GM must be a strong leader to be effective.

MEETINGS AND COMMITTEES

An organizational chart is useful in identifying the formal reporting and authority relationships of a hotel or a lodging enterprise. However, it is not of much help in coordinating administrative units at the department and subdepartment level.

Consider this scenario. If the director of sales increases group bookings without consulting other hotel departments, a disaster is in the making. If the reservations department is not consulted, the sales department might guarantee more rooms to a group than are actually available at a price lower than the reservations department's quarterly goal. If the convention services manager is not consulted, necessary meeting rooms might not be available. If the food and beverage department is not consulted, the group might be sold a banquet that exceeds the hotel's capabilities. If the accounting department is not con sulted, credit terms might be extended that violate the credit manager's policies. If the front office is not consulted, the group might experience lengthy delays at check-in.

The executive operating committee (EOC) of a hotel, made up of the general manager and senior executives from each department, is designed to increase the level of coordination between departments. There is no standard membership for this committee, but it usually comprises those executives who report directly to the general manager.

The EOC is also responsible for a hotel's major budgetary units, such as the food and beverage and housekeeping units. EOC functions depend on how the hotel GM chooses to use the group, the GM's style, and the structure of other management meetings in the hotel. Typically, the EOC meets weekly, focusing on matters ranging from day-today operational issues (daily function schedules and labor and food cost control) to comparing budgets with actual operating results.

Many hotels increase communications through an elaborate structure of additional committees and meetings, including some or all of the following:

operations, staff, sales forecast and marketing, departmental, subdepartmental, credit, safety, energy conservation, and employee meetings.

The operations committee comprises the general manager, department heads, front office manager, manager on duty, and representatives from housekeeping, security, engineering, and food and beverage. This committee might meet four or five times per week for 15 to 20 minutes to review upcoming activities and assess the results of previous activities.

The staff committee might include the GM, department heads, and all subdepartment heads who report to the department heads. This committee, which might meet weekly for one or two hours, reviews the prior week's performance, the current week's activities, the next week's plans, and special projects. The staff committee also presents performance awards to employees.

The sales forecast and marketing committee might meet one to four times per month for several hours so the GM and department heads can review room demand for the coming 90 days and devise strategies to increase room nights and to increase average daily rates by up-selling potential guests to higher-rated rooms with perhaps more amenities or services.

The departmental committee consists of the department head and his or her subdepartment heads, managers, and supervisors. Meeting once or twice per month for an hour or so, the group reviews departmental issues. Similarly, subdepartment committees meet monthly for about an hour so the subdepartment head, managers, and supervisors can address issues unique to their subdepartment, such as the selection of a new type of floor cleaner by the housekeeping department or a more energy-efficient light bulb by the engineering department.

The credit committee includes the general manager, the controller, sales, the front office, reservations, catering, and the credit manager. Meeting monthly for an hour, the committee reviews those guests and clients of the hotel who were granted credit but have not settled their account.

The safety committee typically comprises representatives from human resources, food and beverage, housekeeping, and engineering. Meeting monthly for an hour or so, the committee reviews safety programs and safety records, addresses problems, and discusses the implementation of new safety regulations.

In some full-service hotels, an energy conservation committee includes the chief engineer, resident manager, food and beverage staff, human resource representatives, rooms staff, and housekeeping representatives. The committee typically meets monthly for an hour to discuss strategies and programs for controlling energy costs. Most full-service hotels convene a monthly meeting or at least an annual meeting of all hotel management and employees to review performance and to distribute awards. This event ranges from an hourlong meeting to a company-wide celebration lasting several hours.

THE FUTURE ORGANIZATION OF HOTELS

As new business practices are evolving as fast as our technologies, resistance to change has become a primary cause of business failure. The future success of a hotel will be driven in large part by the ability to foresee and capitalize on change. As we go through global transitions, the successful hotel will examine the key factors that will not only define success but also the ability to survive in coming years. These key organizational trends must be acknowledged by the successful hotel organization: visionary leadership, globalization, diversity, flexibility, flat structure, customer focus, zero defects, network orientation, and being in the information fast lane.

The organization must be able to respond to increasingly globalized sales, the movement to maintaining sales offices in many countries and hotel properties across the globe, and an increasingly globalized labor market. Intercontinental Hotels Group recently introduced a new organization structure to more efficiently use regional and global resources to drive higher levels of innovation, customer focus, and revenues.

Diversity means the organization must respond to a workforce that is heterogeneous sexually, racially, and chronologically; innovation and conflict/ communication issues; and different styles of interaction, dress, presentation, and physical appearance.

Flexibility in the modern hotel organization means assuring that systems, processes, and people can respond differently to different situations; fewer detailed rules and procedures; greater autonomy and encouragement of initiative; customizing employment relationships to include telecommuting and job-sharing; and lifetime employability rather than lifetime employment.

The trend toward flatness in hotel organizations means fewer levels of management, workers empowered to make decisions, and fewer differences in responsibility. The organization's employees need to believe in a sense of entrepreneurship that reacts proactively to market diversity. Traditional organizations that follow well-documented rules must give way to leaders who can balance a sense of discipline with one of flexibility.

If the customer is king or queen in the 21st century, hotel organizaions will be best served by focusing less on their hotel assets as measures of success and more on their customers. This involves a fundamental shift from viewing the real estate asset as the wealth creator to seeing the customer as the key to building shareholder wealth. A customer focus must reflect business decisions at all levels of developing and operating a hotel organization. Pursuing such a course will inevitably impact shareholder wealth.

The Japanese concept of zero defects in products and services can yield tremendous benefits for a hotel business organization. In practical terms, the hotel industry finds it extremely difficult to meet the standard of zero defects in service. Hotel services are based primarily on people, not on computers or other equipment. Twenty years ago, a business executive did not expect a consistent and predictable level of service wherever he or she traveled. Today that is the standard, not the exception, as is the expectation of sophisticated technology in hotel rooms to support business needs. With customer discrimination so acute, it is not surprising that brand loyalty is a diminishing commodity in the hotel industry.

Today's hotel organization must recognize the need for visionary leadership. The old command-and-control model of leadership is giving way to a focus on leadership in ideas, information, inspiration, vision, and teamwork. A failing hotel organization is overmanaged and underled.

The networked hotel organization can facilitate direct communication across unit and property boundaries, ignoring the chain of command; cross-unit team structures; outsourcing and downsizing; strategic alliances with competitors and others; customization; and decentralization.

Being in the information fast lane is critical. The traditional role of information technology as a back office support for accounting and bookkeeping has clearly moved to front and center stage. Information technology today influences all aspects of business from corporate strategies to organizational structure. Technology was once viewed as a way to reduce costs by replacing people. That attitude has been firmly supplanted by the idea of seeking information technology support for the creative work all organizations must pursue. Information technology must enable organizations to react more speedily to market needs and, of course, produce the fulfillment of customer demands both quickly and accurately. To do this it must operate on a decentralized basis. Information technology delivers, but it has to deliver the right information to the right people at the right time.

THE HOTEL STAFFING SYSTEM

Staffing, which is one of a hotel's most important management functions, is an ongoing challenge because of the high rate of employee and manager turnover. Full-service hotels can experience annual turnover rates in excess of 100 percent in certain employee classifications. Some managers consider an annual employee turnover rate of 33 percent low. (In other words, in a single year, one-third of a hotel's employees must be replaced.) At this rate, the entire hotel must be completely restaffed every three years. The higher the turnover rate, the larger the number of employees who must be replaced. For example, if a hotel with 450 employees has a 75 percent annual turnover rate, it will be completely restaffed every 16 months. Staffing is the responsibility of the human resources department.

In an attempt to reduce employee turnover, hotel and lodging businesses are giving increasing attention to job design, seeking to enhance those job characteristics that give the employee the greatest satisfaction and motivation. Good job design must take into account the needs of employees as well as the demands of the job. Well-thought-out job design begins when management conducts a job analysis—that is, a thorough evaluation of the specific tasks performed for a particular job and the time required to perform them. Job analysis is an ongoing process, as many jobs change with improvements in technology and pressure to improve product quality.

The job analysis is the basis for the job description and job specification. A job description includes the job title, pay, a brief statement of duties and procedures, working conditions, and hours. The job specification is an outline of the qualifications necessary for a particular job.

In response to the limits of specialization, organizations can redesign jobs to improve coordination, productivity, and product quality while responding to an employee's needs for learning, challenge, variety, increased responsibility, and achievement. Such job redesign often involves job rotation, the systematic movement of employees from one job to another; job enlargement, an increase in the number of tasks an employee will do in the job; job enrichment, the attempt to give the employee more control over job-related activities; and flextime, a flexible work schedule that permits employee input in establishing work schedules. In team-driven job redesign, a concept similar to job rotation, employees can transfer back and forth among teams that provide different services or products.

Hotels recruit employees from a variety of sources. Newspapers and employee referrals are used to recruit nonskilled hourly employees. Supervisory and management employees generally are recruited through colleges and universities, promotions from within, professional associations, and management recruiters. Hotels that take more time in making their selections are more successful in retaining employees.

Discussions of employee training and development often concentrate on training techniques without giving a full explanation of what a hotel is trying to accomplish. As training and development impart job skills and educate employees, supervisors, and managers, they also improve current and future employee performance, which affects the bottom line. Effective training includes problem solving, problem analysis, quality measurement and feedback, and team building.

Performance evaluation, also called performance appraisal, is the systematic review of the strengths and weaknesses of an employee's performance. The major difficulty in a performance appraisal is quantifying those strengths and weaknesses. The performance of some jobs is easy to quantify, while for others it is more difficult. An important part of the appraisal process is a well-established job description, so that the employee and the supervisor have similar expectations. Compensation includes the monetary and nonmonetary rewards that managers, supervisors, and employees receive for performing their jobs. In order to set compensation levels, the human resources department must periodically conduct job evaluations, which determine the value of the job to the hotel. Knowledge of the value of the job to the organization and of wage rates for each job classification allows the hotel to establish a fair compensation policy.

CAREER PATHS AND OPPORTUNITIES

If you complete your course of study and graduate with a bachelor of science degree in hotel management, most likely you will enter the business at the managerial level. Along the way you will have learned that a successful manager provides clear direction, encourages open communication, coaches and supports people, provides objective recognition, establishes ongoing controls, follows up and gives subordinates feedback, selects the right people to staff the organization, understands the financial implications of decisions, encourages innovation and new ideas, gives subordinates clear-cut decisions when needed, and consistently demonstrates a high level of integrity.

There are three levels of management careers in the hotel or lodging business: first-line, middle, and top. First-line refers to those who have day-to-day contact with the guests and clients of a lodging business. The first-line manager oversees the work of the supervisors and line employees. In a hotel or lodging business, first-line positions may include assistant manager of housekeeping, assistant front office manager, and assistant restaurant manager. First-line managers are responsible for a hotel's basic work, such as checking guests in and out, making up the guests' rooms, and preparing and serving the meals. First-line managers are in daily or near-daily contact with line employees.

Middle management of most hotel or lodging businesses includes the department manager, general manager, and any position between those levels. Depending on the size of the hotel, the regional manager (who supervises the general managers of the hotels in his or her region) can also fall into this category. Unlike first-line managers, those in middle management plan, organize, lead, and control other managers' activities and are responsible for the performance of their departments. Top management comprises a small group of managers such as the chief executive officer, president, or vice president. Top management is responsible for the performance of the entire hotel business as well as for supervision of the middle managers. The top manager is accountable to the owners of the financial resources used by the organization, such as the stockholders or executive board.

As you have already seen, there are numerous attractive careers in the hotel and lodging business.

The following is one of the many paths your career might follow:

- Assistant manager of the reservations department
- Reservations department manager

- Rooms department manager
- Resident manager
- General manager
- Regional manager

In a full-service hotel or lodging business, the movement from entry-level position to general manager might encompass 15 years. Career advancement in a limited-service hotel or lodging business can occur more rapidly. A career in a limited-service hotel or lodging establishment might commence at the assistant general manager level, with movement to general manager within three years and to district or regional manager within five to eight years. This accelerated pace is due in large part to the more restricted range of services the manager must master before advancing.

REVIEW QUESTIONS

- What is the difference between work teams and quality circles?
- Compare centralization and decentralization.
- Which span of control results in a tall organization? A flat organization?
- Should a first-line manager delegate more or less responsibility than a top manager? Explain your answer.
- List the principal functions of each major department of a fullservice hotel.
- A customer notifies the front desk that a table in her room has a broken leg and that when she set her room service tray on it, it tipped over and scattered the food on the floor. List the departments to which this information must be conveyed and the actions they must take.

KITCHEN DESIGN

When you decide to remodel your kitchen, you may want to entirely change its design. Often, remodeling a kitchen is not so much updating appliances, countertops, and cabinets as it is getting to the very core and essence of the kitchen. This means rethinking the plan and flow of the kitchen. Let's look at basic, tried-and-true kitchen design layouts.

THE ONE-WALL KITCHEN DESIGN

Very basic, the one-wall kitchen design is perfect for long and narrow kitchens. It's simple and, as far as kitchen remodels go, fairly inexpensive. While it does not utilize the classic kitchen triangle, its linear design still allows for unimpeded traffic flow. Counter space is at a premium. The one-wall kitchen design is not the best design, though. It can be improved with either the corridor or L-shape kitchen design plans—provided kichen square footage is available.

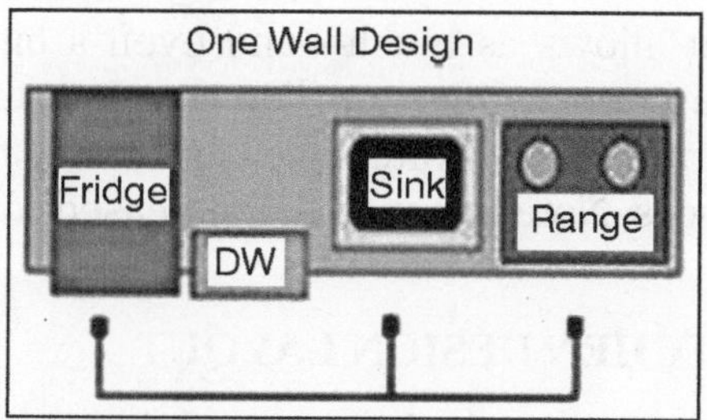

Most likely, everyone has at one point lived in a house or apartment with a corridor kitchen design.

KITCHEN LAYOUT: CORRIDOR STYLE

With counters on both sides, the corridor style kitchen design is highly functional because it uses the classic kitchen triangle. This layout gives a bit more space for counters and cabinets, though crowding between the two main work spaces may be a problem. Still, the corridor-style kitchen design is a highly affordable and functional plan.

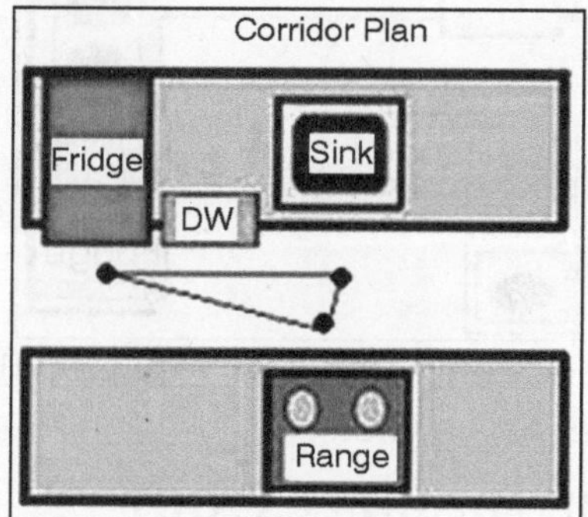

THE L-SHAPED KITCHEN DESIGN PLAN

Along with the corridor plan, the L-shaped kitchen design plan is very popular, though this plan does away with the crowding problems found in the corridor plan. We still find the kitchen triangle, though the walk from range to refrigerator is longer.

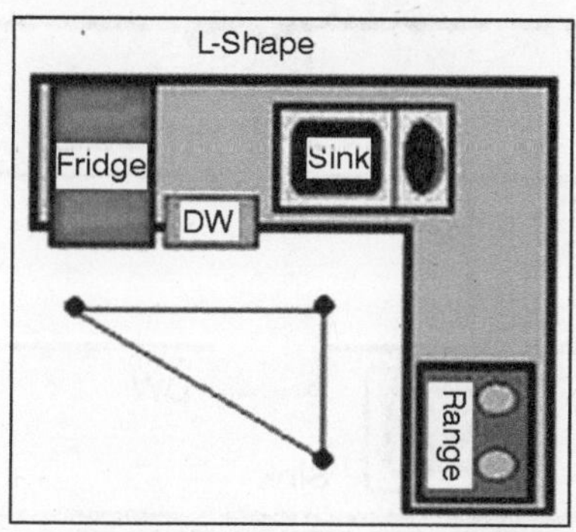

This arrangement allows as much—and even a bit more—counter and cabinet space than the L-shaped design. The corner counter space is difficult to reach for food preparation, and is often used for storing mixers, toaster ovens, and other small appliances. Note that with the increased counter space, a double-sink can installed.

THE DOUBLE-L KITCHEN DESIGN LAYOUT

A highly evolved kitchen design layout, the double-L allows for *two* workstations. The smaller "L" has a cooktop and a second sink. The major cooking operations are focused on this area, while food prep goes on in the larger of the "L" spaces. This larger "L" also has tons of open counter space because the cooktop has been moved away.

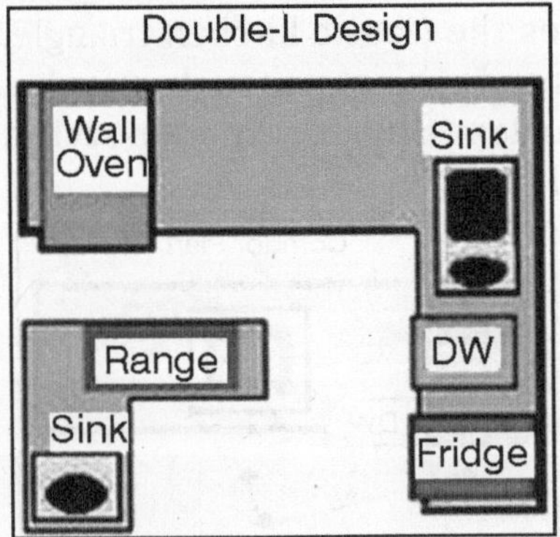

Note that not much space is gained for cabinets: the smaller "L" has only a short run along the wall for cabinets.

THE U-SHAPE KITCHEN DESIGN LAYOUT

The U-shape kitchen design plan can be thought of as a corridor-shape plan—but with a closed end. The closed end gives extra room for a range or a sink. This arrangement maintains good workflow by means of the classic kitchen triangle. The closed end also provides plenty of space for extra cabinets.

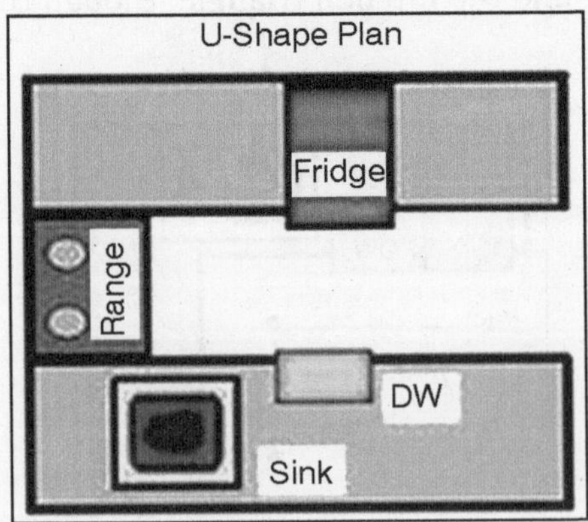

PURCHASE-ITS PROCEDURES

	Amount	Procedure	
	Up to Rs.250 In each cash memo	By local/cash purchase in accordance with rules in force, after recording that purchase has been made after a thorough market search and after ensuring that price has been reasonable.	
	Rs.251 to Rs.500 In each cash memo	Along with a certificate that in view of extreme urgency, it was not possible to effect the purchase by inviting quotations.	
	Rs.501 to Rs.20,000 For each item of purchase/works	(i)	by limited tender – through issuing enquiries to a limited number of suppliers/contractors as approved by the Stores & Purchase Committee – based on the lowest technically acceptable tender/quotation out of at least three valid tenders/quotations.
(ii)		by single tender directly from the manufacturer or sole distributor when the article to be purchased is a proprietary item. For purchase on the basis of single tender for non-proprietary items.	
(iii)		by emergent tender – by obtaining three sealed quotations after making a thorough market search by a committee specially appointed by the Director or D.G. for extremely urgent and strict time-targeted requirements and only in exceptional circumstances when normal issue of enquiries is not permissible because of extreme urgency. All such cases shall require prior approval of Director or D.G.	
	Above Rs.20,000 For each item of purchase/work	(i)	by open tender – through press advertisement, and based on the lowest technically acceptable tender out of at least three valid tenders.
(ii)		by limited tender – as indicated in item 8.2(a) and 8.2(b)	
(iii)		by single tender – directly from the manufacturer or sole distributor when the article to be purchased is a proprietary item. For purchase on the basis of single tender for non-proprietary items, the case shall be referred to Executive Committee or Director General when the amount exceeds Rs.10,000.	

INDENT STAGE

The indent for the procurement of stores should be raised in the prescribed proforma. Separate indents for assets/non-consumables/consumables of different categories should be raised.

The indent or should certify on the indent:

- Project Code
- Purpose of purchase
- Budget sub-head
- Whether the unit cost is below Rs.20,000/-.

The indent must indicate detailed specifications and reliable sources of supply – if known. Indents involving estimated value beyond Rs.20,000/- will require approval by the Director or DG< NCSM. If repeat orders are likely to be placed during next one year depending on possible future requirements, the indent shall clearly indicate the 'quantity immediately required' and also 'additional quantity which may be required in next one year'.

This must be indicated in the enquiry issued to suppliers. For indents relating to the procurement of Stores on "proprietory" basis, the proforma giving technical reasons should be signed as under by the Head of the Division/ Area.

PROFORMA FOR PROPRIETORY /SPECIFIC BRAND GOODS

- Description of item, type, model no. including general specification.
- Name of the manufacturers/suppliers of the item suggested by the Indentor.
- Is there any other manufacturer/supplier of this item in the market.
- If answer is 'yes' for item 3 above, why same is not considered? (Indentor must bring out comparative mandatory advantages of the item selected from the types offered by others).
- If in the knowledge of the indentor, there is no other manufacturer/ supplier or this item, why public/limited tender cannot be issued to locate possibly more suppliers?

Note: Standard products viz. furniture, fans, electrical fixtures, common tools, meters etc. for which comparative products by other manufacturers are available in market, shall not be considered as 'proprietory item'.

In the event of any major change in specifications of an indented item, requiring fresh quotations, a revised indent should be given.

EMERGENT Indents be raised in exceptional circumstances of extremely urgent and strict time-targeted requirement, specifying the need for procurement on emergent basis. Such indents will be processed, after obtaining the approval of the Director or D.G., by a Committee consisting of three Scientific/Technical Officers.

PROCUREMENT OF STORES ON THE DGS&D RATE CONTRACT

Where DGS&D Rate Contract exists, material may be procured at DGS&D rates, provided the suppliers accept supply orders on museum's order forms and agree to supply materials directly to museum/centre instead of channelizsing through DGS&D.

INDENT REGISTER

All purchase indents shall be entered in the Indent Register and its serial number with date may be given on the original copy of the indent and office copy.

RATE/RUNNING CONTRACTS/AGREEMENTS

Rate/Running Contracts/Agreements may be concluded with the manufacturers for the supply of consumables, livery items, stationery and other items of regular use depending upon their total annual consumption and cost. The period of currency of such rate/running contracts/agreement normally should be one year.

SYSTEM OF INVITING QUOTATIONS/TENDERS

If the total estimated value of an item exceeds Rs.20,000/- OPEN TENDERS through press and/or India Trade Journal should be invited. Copies of the tender letters should also be simultaneously mailed to known manufacturers/suppliers in the line.

The condition of press advertisement may be dispensed with in the following cases:

- If the sources of standard suppliers are definitely known and no useful purpose is likely to be served by inviting open tenders and tenders are invited from all such known sources.
- When materials/work are to be of very high standard and quality which can be ensured only by inviting tenders from selected specialized agencies of similar rank and standard; and when press advertisement would bring in incompetent suppliers/agencies who would not be able to execute the work or produce desired quality and would simply under quote. Indentor has to record full justification and obtain prior approval of DG, NCSM before processing such cases. Director or D.G. will ensure that a shortlist of contractors/manufacturers/suppliers of similar standard is prepared for such items after a press advertisement or thorough market survey is made and after all

credentials are scrutinized by a committee appointed by Director or D.G. Care must be taken to issue enquiries only to the parties enlisted in the shortlist and to all of them.

COST OF PUBLIC TENDER DOCUMENT

Estimated value	Cost
Upto Rs.1.00 lakh	Rs. 10/-
Above Rs.1 lakh and upto Rs. 5 lakh	Rs. 20/-
Above Rs. 5 lakh	Rs.100/-

- No Tender Document Fee will be levied in such of the cases, where suppliers will be asked to submit their offers on the basis of the advertised specifications and it is not necessary to issue separate tender documents.
- No Tender Document Fee will also be charged from the Government Departments/Public Sector Undertakings and Consulates of foreign governments in India. Tender documents, if requested may also be supplied free of cost to Small Scale Industries Units borne on the list of National Small Industries Corporation.

EARNEST MONEY

In such of the cases, where fabrication and/or supply/erection/ commissioning of the equipment is required, earnest money @2.5% of the estimated cost of the item of stores rounded up to nearest tens of Rupees, should be obtained from the bidders along with their offers by way of cross demand drafts/bank guarantee from any of the nationalized banks.

Earnest Money Deposits in respect of such of the offers, which have not been accepted will be returned within 30 working days from the date, a final decision is taken about the source from which the equipment is to be procured or within two months from the date of the opening of the tenders.

LIMITED TENDERS

Where the total estimated value of the indent does not exceed Rs.20,000/- limited tenders may be invited from at least five or preferably seven firms in the line in form No.P-3.If the number of the possible sources is less than seven the enquiries shall be mailed to all the known firms in the line.

At least seven days but preferably 14 days should be given to the suppliers to submit their offers, which may be reduced in case of 'Emergent Tenders'.

For indents involving an estimated value below Rs.500/- purchase may be made after comparing rates over phone or by market survey and recording on vouchers that the price is reasonable.

ACCEPTANCE OF LATE/DELAYED TENDERS

Late tenders/delayed tenders are NOT to be considered:

- All tenders in double sealed envelopes will be put in the tender box. The tender box will be kept in the room of the Administrative Officer. The tender box will be kept locked.
- Tenders which are received in the cover bearing no indication on it that there is a tender document inside and opened at Dak stage will be closed back in a new cover along with the original cover (bearing postal stamp, if any) and initiated by Section Officer/Administrative Officer with suitable remarks and indicating tender reference number and due date of its opening on the cover before its being put in the tender box.
- Tenders will be opened in the presence of the tenderers, if any, by three officers of the Purchase & Stores Committee on due date of the opening of tenders. Summary sheet of the tenders received will be compiled in form P-4.
- Tenders must not be opened before their due time and date of opening.

A tender register will be maintained as per form P-5.

EXTENSION OF LAST DATE

Last date for receiving tender may be extended, after obtaining approval from Director or D.G., if less than three tenders are received by due date. Such extension shall be intimated in writing to all parties from whom tenders were invited.

THREE QUOTATIONS/TENDERS

Minimum three valid quotations or tenders are required for comparison of prices before making purchase unless it involves proprietory item. In case of extreme urgency, purchase may be made on the basis of single tender up to Rs.10,000/- and two tenders up to R.50,000/-, if three tenders are not received by due date, provided that the Director records in file that I) loss of time due to extension will severely affect the progress of a time-targeted work, and ii) the price quoted in tenders are reasonable and competitive.

Such cases shall be exceptional and shall not be a routine matter. For single tender over Rs.10,000/ or for purchase over Rs.50,000/- the case shall be referred to Executive Committee or D.G., NCSM.

SPECIAL CLAUSE FOR DRAFTS

A special clause shall be inserted in the enquiry that all payments shall be made by account payee cheques. If contractor/supplier insists on payment by bank draft, the bank commission shall be deducted from the bill.

COMPARATIVE STATEMENT

Comparative Statement should be compiled for all accepted offers, indicating net price of uniform units, i.e. after taking out the element of discount, if any, brand make offered, details of all levies like excise duty, sales tax, surcharge, delivery charge etc., delivery schedule, terms of supply payment guarantee and other relevant particulars. Comparative statement will be compiled in form P-6. The comparative statement shall first be checked by Stores & Purchase Officer and then placed before the Indenting Officer.

RECOMMENDATIONS

Recommendations on offers shall be made by the Originators of the indent. In case where the offer of other than the lowest bidder is proposed to be accepted, such recommendations should be duly confirmed by the Director or D.G.

Technical reasons for by-passing the lowest offer should be recorded specifically.In cases where the proposal is for purchase of standard/shelf items, such as office equipment, air-conditioners, furniture, tyres and tubes etc., the Purchase Officer may himself finalise the recommendations on the basis of the lowest offers where such offers are precisely identical to the indentor's requirement without reference to the indentor.

In case of any variation or in case of doubt, the offers may be referred to the indentor.

After getting the indentor's recommendations, the Purchase Division will process the case further and obtain sanction of the Competent Authority in the prescribed proforma.

PRE-AUDIT

In all cases pre-audit is necessary before approaching the sanctioning authority:

- Availability of budget provision has to be ensured from Finance Section before approaching the Sanctioning Authority.
- Finance & Accounts Officer shall ensure that all requirements of purchase procedure have been observed. In case of any deviation, approval of Director or D.G. must be obtained.
- It should be checked whether the lowest offer is accepted and in case the lowest offer is not accepted, reasons for the same are recorded in writing by the authority to decide the case and the same is available in file.
- It should be verified whether in the Purchase Order there are any unusual conditions or deviations from the prescribed standard terms and conditions and if so whether the approval of the Competent Authority has been obtained to accept the same.

- Wherever there is disagreement by F.A.O., the matter will be referred to the Director or D.G. and his decision will be final.

SANCTIONING AUTHORITIES

(i)	Beyond Rs.25 lakh for purchase or beyond Rs.50 lakh for works	Executive Council of Museum/Centre or Governing Body, NCSM
(ii)	Rs.50,000 to Rs.25 lakh for purchase and Rs.50 lakh for works	Executive Council of museum/centre or D.G., NCSM
(iii)	Upto Rs.50,000	Director of museum/centre
(iv)	Up to Rs.10,000	Certain Officers who are specially authorized by DG, NCSM
(v)	Up to Rs.1,000	Project Coordinators
(vi)	Items of Appendix B of NCSM Byelaws	Administrative Officer or such other Officer as authorized by D.G., NCSM

Note:

- Sanctioning power vests on such authorities subject to budget provisions.
- All purchase proposals must be sanctioned by the appropriate authority before placing the order or making the cash purchase.

EX-POST-FACTO SANCTIONS SHALL BE AVOIDED BY ALL MEANSSIGNING OR PURCHASE ORDRS

All Purchase Orders will be signed by the Stores & Purchase Officer or in his absence by Section Officer/Administrative Officer.

Purchase Orders will be issued in the proforma as under:

- Orders for indigenous Stores in form P-8.
- Orders from imported Stores in form P-9.

All purchase made on credit basis must be on the basis of the firm Supply Orders. No Supply Order shall be issued independent of the Purchase Department. This procedure will apply to even such stores, which are expected to be received free of cost.

All Purchase Orders, before their issue, will be entered in the Indian Order (commitment) Register and Foreign Order Register. The serial number of the Order Registers will always be indicated after the file numbers. Serial numbers of Foreign Order Register may be given after the file number with prefix "F".

Copies of purchase order shall be endorsed to:

- Stores.
- Indentor.
- Finance.

ESCALATION OF PRICE

No escalation of price is normally admissible after opening the tender or after placing the order.

For cases, where escalation is apparently justified because of statutory levies after opening of tender or placement of order, the matter must be resubmitted to F.A.O., and then Director or D.G. for scrutiny and decision.

ISSUE OF REPEAT ORDERS

Repeat Orders may be placed on the following conditions:

- Within six months from the date of original order. This may be extended to one year on discretion of Director or D.G.
- The items to be ordered again should be of identical specifications.
- The Purchase Officer should be satisfied that there is no downward trend in the market price.
- The quantity to be ordered shall not normally exceed the quantity in the original order. Director or D.G. may consider repeat order for larger quantities in case of extreme urgency.
- The original order should have been placed for a proprietory item or on the basis of lowest technically acceptable quotation.
- The original order should not be an 'emergency' order or an order placed on 'Preferred Delivery' basis.
- Efforts should be made to obtain quantity discount from the supplier.
- Repeat Order may also be placed on the basis of the Purchases finalized by a sister museum/centre.
- Several repeat orders may be placed during one year after placing first order, depending on time to time requirements, if such possibility is indicated in the enquiry and the contractors/suppliers have been asked to quote their rates accordingly.

FOLLOW UP OF PURCHASE ORDERS

Purchase Section will be responsible to ensure that Purchase Orders are regularly followed up and materials ordered are supplied by parties concerned within the stipulated delivery period.

Orders follow-up letters will be regularly issued to pursue suppliers to supply materials as per delivery schedule.

When the delivery date have expired, the firm should be asked to indicate the revised date by which they expect to supply.

The extension in deliveries may be granted subject to penalties in the contract, if any.

PASSING AND PAYMENT OF BILLS

Since prompt payment of the amount due to the suppliers builds the credit worthiness of the museums/centers in the market, all concerned viz. the Purchase, Stores,

Accounts Department and the Indentor shall ensure that the payment of all bills are arranged as expeditiously as possible.

Before passing bills for the supply of stores, it shall be ensured:

- That the sanction of the competent authority to the expenditure exists.
- That stock entry and inspection certificate has been recorded on the body of the bill by the Stores Officer.
- Rate and other charges are in accordance with the Purchase Order.

The bills shall be submitted by the supplier in triplicate. These shall be received in the Purchase Section and entered into the "Register for bills and payments" and action initiated by forwarding all the three copies to the Stores Department for inspection and recording stock entry certificate thereon in respect of quality and quantity of goods received.

On receiving the materials in store and even before receiving the suppliers bill, the Store Supervisor/Keeper shall immediately inform the indentor, over phone, of the arrival of materials in store.

The indentor shall promptly inspect materials or arrange inspection by a Competent Officer and give a certificate on the reverse of the bill in the manner.

FOR PURCHASE OF STORES

Certified that stores have been examined by me and found acceptable in accordance with the quality, quantity and specifications ordered for.

Inspecting Officer

FOR CASH PURCHASE ONLY

Certified that the materials were purchased at reasonable local market price/directly from the manufacturer. It is also certified that stores have been examined by me and found acceptable in accordance with the quality, quantity and specification ordered for.

Inspecting Officer

FOR EXECUTING SOME WORK

Certified that the work has been done satisfactorily and as per our specifications and requirement. The claim is reasonable.

Inspecting Officer

In case some materials are found defective, it shall be recorded in the inspection certificate. The Store Supervisor/Keeper shall immediately contact the party over phone or through mail and ask for prompt replace-ment, after which he shall again arrange inspection by the indentor.

If the replacement is expected to take time, the Store Supervisor/Keeper shall give stock entry certificate for the portion of order received in good condition and pass on the bill to the Purchase Section for part payment, if so recommended by the indentor.

The Store Supervisor/Keeper will then enter the materials in stock and give the following certificate on the bills in triplicate:

"Certified that the material has been taken on stock vide Daily Receipt Register No........... Sl. No. ...... dt. .... & Stock Ledger page/ Cardex Card No. ...."

Stores Assistant **Stores Officer**
Date: *Date*:

The 1st and 2nd copies of the bill shall be returned to the Purchase Department along with certification both from the Stores Department and the Indentor.

The details of freight or demurrage charges incurred or other deductions to be made, if any, must also be detailed in the certificate recorded by the Stores Department. The third copy shall be retained in the Stores Department.

The Purchase Department shall after scrutiny and where necessary, making deductions for the rejected stores or other reasons record the following certificate on the body of the bill and shall forward the same to the Accounts Department for arranging payment.

"Certified that the bill is in order and may be admitted for payment for Rs........... Budget sub-head .... Project Code No. .... Sanction O.M. No. .... dated .... for Rs. ...."

Purchase Officer

Suitable entry shall also be made in the indent register and in the Indian/ Foreign Order Register as well as in the "BILL PASSING REGISTER". The original copy of the bill shall be forwarded to the Finance & Accounts Officer on the above form for audit and payment and the duplicate copy shall be retained in Purchase Department.

No file will ordinarily be sent to the Accounts Department along with the bill unless demanded by them in a particular case, in view of the fact that the Purchase has been pre-audited before the sanction was obtained and a copy of the sanction memo and order are already available in Accounts Department.

The Finance Section shall deduct the bank commission from the bill if the supplier/contractor insists on payment by bank draft. Otherwise payment shall be made by account payee cheques. The Finance and Accounts Officer after making payment of the bills, shall return the covering memo i.e. Form P-13 to Purchase Department for record in the relevant file mentioning retrenchment or observation, if any, therein.

The Purchase Department shall on receipt of discrepancy/rejection report from the Stores Officer take steps for the replacement or recovery of cost of materials found damaged or received short from the suppliers, or from the Insurance Co., Railway or others as the case may be.

All cases involving recovery/refund of claims shall be entered in the Register of Claims and Refunds All cases, in which advance payments have been made against purchase orders before the receipt of stores, should be entered in "Advance Register".

In addition, where letters of credit have been established, their particulars should be entered in the L/C Register. These cases may be carefully and vigorously pursued till their finalisation and adjustment. Requisite entry against adjustment of advance payments and finalisation of L/C cases will be made against the relevant entries in the Advance Register and L/C Register.

ALL BILLS MUST BE SETTLED LATEST WITHIN 15 DAYS from the date of the receipt of bill from the party. Different Officers must follow a strict time target for passing the bills according to the following schedule:

Entry in Bill Register by Purchase Section	:	1 to 2 days
Inspection Certificate by Indentor	:	2 to 3 days
Stock Entry by Stores	:	2 to 3 days
Checking by Purchase	:	2 to 3 days
Checking by Finance & issue of cheques/drafts	:	2 to 5 days

TOTAL 8 to15days

All sanctions must ensure that bills are promptly sent to the next section through suitable peon books so that track can be kept.

However, where some commitment for settlement of the suppliers bill within some specified period has been made, efforts by all concerned will be made to ensure that payment is arranged within the agreed period.

PURCHASE IN RSC/DSC

For purchase in Regional/District Science Centres having no Stores & Purchase Officer nor a Finance & Accounts Officer, the following procedure shall be adopted:

PROCEDURE OF PURCHASE AND SANCTIONING AUTHORITY

Amount of Purchase	Procedure	Sanctioning Authority
Up to Rs.250	Cash purchase	Project Coordinator
Rs.251-500	Cash purchase as a very special case after recording reasons for urgency	Project Coordinator
Rs.501-1000	On the basis of lowest quotation out of at least 3 quotations	Project Coordinator
Rs.1001-10000	On the basis of lowest quotation out of at least 3 quotations	Such Officers who are au-thorized by DG to accord sanction upto Rs.10,000/-
Rs.10001-20000	On the basis of lowest tender out of at least 3 tenders (limited tender)	Director of respective museum/centre or DG, NCSM
Rs.20001-50000	On the basis of lowest tender out of at least 3 tenders invited on open tender basis through press advertisement	Director of respective museum/centre or DG, NCSM
Above Rs.50000	Refer to Executive Committee or DG, NCSM	-
Note:	All purchase shall be effected only after obtaining prior sanction from the competent authority. Proposals for ex-post-facto sanction shall not be admitted without DG's approval.	

NO HAND/VERBAL QUOTATIONS

For purchase over Rs.500 and upto Rs.20,000 enquiries must be issued to known suppliers in approved format giving reasonable time for submission of quotations. No purchase shall be effected through hand quotations or verbal quotations unless specially authorized by DG, NCSM on ground of extreme urgency prior to issue of such enquiries.

LIMITED TENDERS

All purchase above Rs.20,000 shall normally be made through open tender through press advertisement.

Purchase may be effected through limited tender only in the following cases:

- Proprietory item.
- When all the names of manufacturers are definitely known and enquiries are issued to all of them (for cars, lifts, computers etc.).
- Emergent purchase, when reason for emergency is clearly recorded in file and purchase is effected by obtaining at least 3 sealed tenders after making a thorough market search by a committee appointed by Director or DG, NCSM.

SHORTLISTING

In specific cases where desired quality of work cannot be achieved through open tender because of underquoting by incompetent organizations such as in civil construction, printing of very prestigious publications, fabrication of exhibits etc., a shortlisting of manufacturers/ suppliers/contractors of comparable standard may be made in advance by a committee appointed by Director or DG, NCSM and purchase shall be effected after enquiries are sent to all such parties included in the respective shortlist. Care must be taken that enquiry is not sent to any stray party outside the shortlist or no party in the shortlist is ignored.

PROPRIETORY ITEM

Proprietory items are those for which alternative products of other manufacturers/suppliers are not technically acceptable and sufficient reasons are recorded for such non-acceptability. Furniture, fans, electrical fixtures, common tools, meters etc. are not proprietory items. Practice of effecting purchase directly from a particular manufacturer, where alternative products are available in market, shall stop forthwith.

CASH MEMO/VOUCHERS

All cash memos and vouchers must be clearly written with indelible ink and bear signature with date. Such cash memos/vouchers must indicate S.T. registration number if S.T. is charged.

SALES TAX

Sales Tax is a statutory levy. It is an essential item of expenditure to be incurred on all purchases taking place in India as NCSM is not exempted from its payment. The enquiry must specifically mention that sales tax, if payable, must be clearly indicated in the quotation by suppliers along with their sales tax registration number. The amount of sales tax shall be inclu-ded in the cost of material to arrive at the nett cost in the comparative statement. In case, it is not admissible at a later stage. Sales tax is required to be paid according to the rates laid down by each State and/or Central Government. Special care shall be taken to ensure that sales tax registration number is indica-ted in the cash memo or bill submitted by the party.

OCTROI/TERMINAL TAX ETC.

The NCSM museums/centers are generally exempted from octroi or terminal tax but the sanction of the exemption in this behalf rest on the concerned local authorities. Appropriate action shall be taken sufficiently in advance to obtain octroi exemption from the concerned authority.

DEMURRAGE/WHARFAGE

No charges for demurrage/wharfage shall ordinarily be incurred in clearing goods at the docks or at the railway terminal. Every effort must be made to avoid this wasteful expenditure.

Whether it is due to the negligence on the part of staff or supplier, responsibility for the same should be fixed and necessary recoveries effected as the case may be. Only in cases where such payment was necessary for circumstances beyond control, sanction or the competent authority shall be taken.

METHODS

You must be going to the market to buy a variety of foods. Some of the foods which you buy like tomatoes, cucumber and fruits are best eaten raw. But others like wheat, rice, dals, potatoes and some vegetables must be cooked before they are eaten.

Do you know why we must cook foods? Because these foods can not be eaten raw. If you eat them raw they will not be tasting well, you will not get all the nutrients present in them and they can also affect your digestive system. Hence you must know different methods of cooking different foods.

But do you know, if the food is not cooked the right ways you will loose many of the nutrients. Therefore, you should use the correct method of cooking to keep all the nutrients safe.

You must also learn to improve the quantity and quality of nutrients of your food. In this lesson you will learn about good cooking and the various methods of improving the nutrient content of food.

OBJECTIVES

After reading this lesson you will be able to:

- Explain the need to cook food.
- Describe the different methods of cooking.
- Explain importance of conserving nutrients while cooking.
- Suggest measures to minimize nutrient losses during preparation and cooking food.
- Describe the methods used to enhance the nutritive value of food.
- Suggest ways to use left over cooked foods.

WHY DO YOU COOK FOOD?

COOKING MAKES FOOD EASY TO DIGEST

When food is cooked it becomes soft so that it is easily chewed and swallowed. The juices that digest food are able to mix well with the softened food. Hence the food gets simplified for use by your body.

COOKING IMPROVES THE APPEARANCE, TEXTURE, COLOUR, FLAVOUR, AND TASTE OF THE FOOD

Cooking improves the taste and flavour of food, changes its colour and appearance. When you cook meat or potatoes etc. you will find that the food has better taste. Cooking meat improves its taste, flavour and colour. Addition of spices and condiments during cooking further improves the acceptability and palatability of foods. You must observe the changes that the foods undergo when you cook next.

WHEN FOODS ARE COOKED YOU CAN MAKE A VARIETY OF DISHES

Cooked foods provide variety in your meals. You must have eaten potatoes cooked in different ways - such as pakora, potato chat, potato parantha, potato vegetables and potato chips etc. Can you list a few food items that can be made with atta? Yes, parantha, puree, chapati, bread, mathi, etc. So, you see that you can make variety of dishes with the same food.

COOKING HELPS TO KEEP THE FOOD LONGER

Do you know why we boil milk? Yes, if you do not boil milk it will get spoiled soon. Boiling milk helps in killing of spoilage organism and makes it last longer. Atta dough gets spoilt after some time. You must have noticed that chapaties can be kept longer than the dough. Can you name a few more foods that will last longer after cooking?

COOKING MAKES THE FOOD SAFE AND STERILE

Raw foods get spoilt because of the harmful micro-organism present in them. These micro-organism get destroyed, when you cook food. Killing germs by cooking makes the food sterile and safe for eating. Milk often contains bacteria that cause tuberculosis.

Do you know what happens when you boil milk? Yes, the bacteria get killed and milk becomes safe for drinking.

METHODS OF COOKING

You often say boil the rice, fry pakoras or bake a cake. What are boiling, frying and baking? These are the methods of cooking.

Food can be cooked:

- By moist heat
- By dry heat.
- By frying in ghee or oil.

COOKING BY MOIST HEAT

In this method water is heated or boiled. The food is put into this boiling water or cooked in the steam which comes out from the boiling water. There are three ways by which you cook food by moist heat.

These are:

- Boiling
- Simmering or stewing
- Steaming

Boiling

In this method food is covered with an adequate quantity of water and heated to a boiling. For example we boil potatoes, eggs, a number of vegetables, rice etc. Usually green leafy vegetables such as cabbage, methi, and spinach are cooked with no water. Whereas vegetables such as green peas, green beans, are cooked with little water. Cereals such as rice and pulses such as dals, legumes, grams are boiled in large amounts of water. There are a few points which you should keep in mind while boiling foods.

- Before boiling, wash the food stuffs thoroughly.
- Cover the food with an adequate quantity of water.
- First boil the water and then put the food.
- Cook in a pan which has a well fitting lid. This way the steam from boiling water will not go out from the pan and the water will not dry up. Food gets boiled faster when the cooking pan has a lid on.
- Do not boil foods longer than needed. Once they are soft and tender, take them off the fire. If food is cooked for a very long time it looses its colour, shape and taste. Potatoes and other root vegetables should be boiled with their skins on.
- Water used for boiling should cover the food. Water soluble nutrients present in foods dissolve in water in which the food is being boiled. If you throw this water, nutrients will be lost. What can you do to save these nutrients? You can use this nutrient rich water to make gravy for another vegetable.

Siimmering or Stewing

Stewing is cooking for a long time in water, below the boiling point. In this method you cook food in a small quantity of water. Once the boiling starts the flame is lowered and the food is allowed to cook slowly.

- *Suitability*: This method is suitable for cooking hard and tough foods like dals, meat and dried vegetables.
- *Advantages*: While cooking by this method we do not have to constantly keep an eye on food. And there are lesser chances of food getting burnt.

Steaming

Do you know what happens when the water boils? Yes, it gives off steam. When food is cooked in water vapour with or without pressure it is said to be steamed and this method of cooking is called steaming. Can you name some steamed foods that you have eaten? Yes, Idli and Dhokla. Steaming can be done for solid and semi-solid foods. Water is heated in a pan on fire. The pan is covered with a clean muslin cloth. Food is placed on the cloth. The steam passes around the food and cooks the food placed above. When you are making idlis, the batter is put in the idli mould, which is then lowered into a container with water at the bottom. Once again it is the steam going around the moulds that cooks the idlis.

- *Suitability*: This method is suitable for making Dhokla, caramel custard etc.
- *Advantages*: Steaming shortens the duration of cooking and helps to conserve nutritive value, colour, flavour and palatability of food. Steamed food is light nutritious and easy to digest. Such foods are good for old people and little children.

Pressure Cooking

Pressure cookers are generally made of an Aluminium alloy which is very strong. Now a days stainless steel cookers are also available in the market. But these are quite expensive. Raw food is put in the container along with water and cooked under pressure. Under pressure the temperature of water is increased up to 1120C.

- *Advantages*: Pressure cooking kills all bacteria and hence the food is safe and hygienic for you to eat. Rice, dal, meat, potatoes, roots, beans, and peas and peas are cooked in the pressure cooker. The food gets cooked faster than boiling. *You can call pressure cooking high temperature short time cooking.*

COOKING BY DRY HEAT

What do we normally eat for breakfast? Some times we eat chapaties, paranthas, purees and some times bread. We also eat rusks and buns. Do you know how these are cooked? Yes, they are cooked by dry heat. Cooking food by dry heat means using hot air to cook the food. There are three methods of cooking food by dry heat using hot air.

- Baking
- Roasting
- Grilling
- *Baking*: Baking is the method in which food is placed inside a closed box called an oven. The air inside the oven is made hot by fire or electricity. The food gets cooked by hot air. Have you seen a bakery in your village or neighbouhood? You must have also seen the big ovens heated by fire in which biscuits, breads and pastries are made in these bakeries. These ovens are also known as 'bhattis'. It is in these ovens that the food is cooked. In the very big bakeries, the air is heated by electricity.You can easily make an oven at home to bake foods. Take an empty oil pin. Put a layer of sand in it and fit it with a lid. Heat this over coal, kerosence or a gas stove. Once it becomes hot, put the food inside and close the lid. Place the tin on a low fire.

Bake food till it is light brown in colour. Do not open the lid very often because the hot air from inside will go out and make the food dry and hard.

- *Roasting*: Another method of cooking food by dry heat is called roasting. Roasting is cooking on a glowing fire. While roasting, the food is put directly on the hot tava, hot stand or hot fire and cooked. For eg. channas, brinjals, potatoes, maize, ground nuts, cashew nuts, papad, meat etc. are cooked by this method. You must, have had chicken or paneer tikkas cooked in this way.
- Grilling Grilling is cooking over a glowing fire. The food is supported on a iron grid over the fire, or between electrically heated grill bars. The grill bars are brushed with oil to prevent food sticking and can be heated by charcoal, coke, gas or electricity. The food is cooked on both sides to give the distinctive flavour of grilling. Potato, sweet potato.

FRYING

Frying is the process of cooking food in hot fat or oil.

Food can be fried in two ways:

- Shallow frying
- Deep frying

Advantages of frying are:

- Fried food is very appetizing
- Quick method of cooking
- The keeping quality of food is increased.

Some precautions while frying food:

- Food should be cut in suitable size and shape.
- Do not put in too many pieces of food at the same time. It will lower the temperature.
- Food should be fried to golden brown colour on both sides by turning over the food if necessary.
  - *Shallow frying*: When you make paranthas or an omellette, you need very little oil for frying. You can fry them on a tava or a frying pan. The food is turned over so that both sides may be browned and cooked. This method is called shallow frying.
  - *Deep frying*: Do you know how to fry foods? Yes, ghee/oil in a karahi is heated to the smoke point. Carefully put the food to be fired in the hot ghee/ oil. The food should fully dip in ghee/ oil. Put only a few pieces of food to be fried at a time. Avoid using large quantity of oil/ ghee or overheating. If some ghee is left over after frying, drain and store in a closed container to be used again. Frying is a quick method of cooking as compared to boiling or stewing.

ACTIVITY

Observe dal, rice, potato curry and brinjal bhartha before and after they have been cooked. Record your observations in the table given below.

Food	Colour		Textrure	Flavour
	Before Cooking	After Cooking		
Dal				
Palak Pakora				
Potato curry				
Brinjal Bharta				

Nutrients Lost During Cooking

So far we have only read about the importance of cooking. Do you know that some nutrients are lost during cooking? Even when we cut and wash the foods, some nutrients are destroyed. Let us understand how some of the nutrients are lost during cooking.

- *Vitamin A*: Vitamin A is found in foods like spinach, methi, carrots etc.When we cook these vegetables not much of vitamin A is lost, but when we fry these foods, like when we make palak pakoras, vitamin A gets destroyed. Can you tell why? If you remember, vitamin A dissolves easily in fats and oils. So, when you fry such foods in oil, vitamin A comes out from the food and goes into oil.
- *Vitamin B*: How is rice cooked? First of all you clean rice and then wash it. Vitamin B being water soluble, goes out of the rice and washes away with the water. If you wash rice by rubbing you are letting more vitamin B wash away. After washing rice, it is soaked in water. Some more vitamin B goes out from the rice into the water during this process. Next, rice is boiled in water. If you use a lot of water to cook rice and throw away the extra water some dissolved vitamin B also goes out with this water. Sometimes, you add cooking soda to soften foods like rajmah and channas. Cooking soda also destroys vitamin B.
- *Vitamin C*: Vitamin C is an important nutrient which is easily destroyed by cooking. When you cut vegetables and fruits rich in vitamin C, some of it is lost. Vitamin C is also lost when you wash vegetables and fruits after cutting and exposing cut vegetables to air for long periods before cooking. When the foods are cooked for a long time or when you throw away water in which you cook them, you loose Vitamin C. You also loose vitamin C when you add cooking soda to the foods. Therefore, cooking procedures that minimize the loss of vitamin C result in conserving all other nutrients.

- *Proteins*: All proteins present in the foods coagulated by heat. Cooking results in softening of proteins in foods such as egg, fish, and meat, becomes water is bound in the process of coagulation. If the coagulated protein is further heated, it loses moisture and becomes dry and rubbery. They also become difficult to digest.
- *Oils and Fats*: When food containing fat are heated, the fat has the tendency to separate from the food. You must have seen that heating milk results in the fat layer floating on top. Oils and fats are used as a cooking medium. Some fat is absorbed during frying. Therefore fried foods such as pakoras, etc. gives us more energy than the boiled foods. When fats, ghee and oil, are heated for long periods of time over and over again, to fry pakoras or purees etc. its quality becomes poor.
- *Minerals*: Minerals like sodium, potassium, etc. dissolve in water. Minerals get lost when food is first cut, then washed and the extra water in which they are boiled, is thrown away.

Conservation of Nutrients

Now you all know that some nutrients are lost when foods are cooked. Nutrients like vitamin B and C are lost when foods are boiled or soaked in water and the water is thrown away. Nutrients like vitamin A are lost when fats are used for cooking foods. If all these nutrients will be destroyed, how will our body get energy to do work, repair body tissues and fight disease germs?

Therefore, you must think of ways of saving these nutrients. Saving nutrients during the process of cooking is called conservation.

- Wash vegetables before cutting them so that minerals and vitamins are not destroyed. Do not wash the foods more than necessary.
- Peel vegetables thinly as vitamins and minerals are found just under the skin.
- Cut vegetables into large pieces just before cooking. Small pieces mean greater loss of nutrients.
- If vegetables are to be cooked in water, put them into boiling water.
- Scrape the peels very thin. Use dry tori or jute to remove peels from potatoes, etc.
- Use just enough water for cooking. Do not throw away the extra water. Use this extra water to cook some other food.
- Do not use cooking soda. Use of tamarind or lemon juice helps to conserve the vitamins.
- Cook rice in just enough water which gets absorbed during cooking? Do you know how much water should be used for one katori rice? Yes, two katori water.

- Cook in a pan which has a well fitting lid. When you cook in an uncovered pan most nutrients are lost.
- Do not overcook the food as many nutrients will be destroyed.

Enchancing Nutrients Content of Food

You are already familiar with the different ways that help you to conserve nutrients during cooking. It would be so nice if we could increase the nutritive value of foods without increasing the cost. Can you suggest some ways of doing so? Let us find out how to do it.

Definition of Enrichment

The process of improving the nutrients in foods by special methods is called Enrichment.

Importance of enhancing nutritive value of food:

- To meet the nutritional requirements of the body.
- To make proper selection and preparation of foods.
- To consume food in a balanced manner.
- To improve the flavour and texture of the food.
- To get variety in food.
- To assist in planning the daily menu, keeping in view the nutrient content of the food.
- To prevent deficiency diseases in the body.
- To develop good food habits.

Methods of Enrichment of Nutrients

There are three methods by which you can enhance or increase the nutrients present in your food.

- Combination
- Fermentation
- Germination

(a) *Combination*:

We all eat combination of variety of foods. For example, you eat a dal or channas etc. with vegetables, salad, curd and chapati or rice. Chapaties or rice will give you carbohydrates, dal and curd will give you proteins, vegetables and salad will give you vitamins and minerals. Combining of foods from different food groups is the easiest way of eating all nutrients.

You can also mix a number of foods in one dish and get all the nutrients from it. Such a combination of foods improves the quality of nutrients.

When you eat rice and dal together you get better nutrients. Do you know why? Cereals lack certain amino acids. And these are present in dals. On the other hand dals lack some other amino acids that are present in cereals. Do you know when you eat dal and rice together, the quality of proteins becomes as good as that of milk. Do you remember the food group to which the carrots belong to? What are carrots rich in? Vitamin A. What will happen to your khichri if you add carrots to it? And what will happen if you eat this khichri with curd?

The khichri will also become rich in vitamin A, calcium and proteins (curd). Can you name some other foods that can be added to your dal rice preparations? Yes, methi, peas, beans, ghee, etc. Can you name some other rice and dal dishes that your family eats? Yes, Paushtik roti, Kachauri, Idli, Dosa, Sambhar, mixed dal, etc. The combination of a variety of foods ensures better availability of nutrients. Combination is the process of combining cheaper and commonly available foods from different food groups to improve the quality of nutrients.

Combination helps you to:

- Eat a diet that has good quality nutrients.
- Use cheaper and easily available foods that enhance the nutrient content of food considerably.
- Provide balanced diet to your family.

(b) *Fermentation*:

Have you ever made bhuturas? These are made by mixing a little curd in maida which is kneaded into a dough and kept covered for few hours. In these few hours the dough rises. Do you know why? When you add curd to maida you introduce micro-organisms which begin to grow at a very fast rate. They start a process called fermentation. Fermentation makes the dough rise and become almost double in quantity. During fermentation the micro-organisms use up some of the nutrients present in the atta and change them into other better quality nutrients. They also make some new nutrients.

Definition of Fermentation: Fermentation is a process in which some micro-organisms are added to the food. They change nutrients already present in the foods into simpler and better forms and also make other new nutrients. Can you name some fermented foods? Curd, bread, khaman-dhokla, idli, etc. are all examples of fermented foods.

Advantages Of Fermentation:

- Fermentation improves the digestibility of foods. The micro-organisms which cause fermentation break the proteins and carbohydrates into smaller parts, which are easier to digest.
- During fermentation of cereals and foods like peas, beans etc. the minerals, calcium, phosphorus, and iron are changed into better quality ones. These are then easily absorbed by the body.
- Fermented foods become spongy and soft and are liked by children and adults.

(c) *Germination*:

Take some whole 'moong' or 'channa' and soak them overnight in a small quantity of water. What do you see the next day? Yes, they become big in size and soft to touch. Now if you tie the soaked dal in a wet cloth and keep for another 12 to 24 hours, you will notice that small, white shoots have started growing from these dals. This process is called germination or sprouting.

Definition of Germination

Germination is a process in which small shoots come out of the dal or cereal when these are kept with small amount of water. The grains and pulses to be sprouted need to be soaked in just enough water so that all of it is absorbed. If the extra water in which they are soaked is thrown away, you will be loosing a lot of nutrients. Grains like wheat, bajra, jawar, etc. can also be sprouted. These grains can then be dried in shade and roasted lightly on a tava. They can be ground and used in many dishes. Pulses like moong, peas, kala chana etc. are also sprouted first and then steamed and eaten after adding salt, chilli powder, lemon juice etc.

The time and water which each grain or pulse needs for soaking and sprouting is different. Normally 8-16 hours are needed for soaking and 12-24 hours for sprouting. The cloth in which the soaked dal is tied should be kept moist all the time. In the winter months sprouting can be done faster by using warm water. If you sprout wheat you can grind it into a fine powder after sprouting. This can then be fermented and bathuras made from the dough. Such a dough will be rich in vitamins. Here the food is first sprouted then fermented.

Germination helps you to:

- Increase the digestibility of foods. Do you know why?
  - Some carbohydrates and proteins are broken down into smaller and easily digestible forms.
  - Grains and pulses become soft after sprouting, so they take less time for cooking and are easy for you to digest.
- Increase the nutritive value of food with no additional cost. Some vitamins and minerals become more when foods are germinated Vitamin B becomes almost double in quantity while vitamin C increases almost 10 times.

When you soak pulses like rajmah, soyabeans etc. in water for a few hours before cooking, it helps to increase their vitamins content.

METHODS OF MIXING FOOD

STIRRING

Food is stirred by a rotary motion of the arm. The purpose of stirring is to mix thoroughly all ingredients.

BEATING

Food is beaten when the motion in mixing brings the contents at the bottom of the bowl to the top and there is a continual turning over and over of a considerable part of the contents of the bowl. The purpose of beating is to enclose a large amount of air.

FOLDING IN

Two foods are blended by putting the spoon or egg-whip vertically down through the foods, turning it under the mass, and bringing it vertically up. This process is repeated until the mixing is complete. The purpose of folding in is to prevent the escape of air or gases that have already been introduced into the mixture.

CUTTING IN

A process used to blend fat with flour. It consists of cutting the fat into the flour with a knife or two knives until it is distributed in as small particles as desired.

CREAMING

A rubbing together of fat and sugar, or a pressing and beating of fat to soften it.

KNEADING

A stretching motion applied to dough when more flour is to be added than can be either stirred or beaten into the mixture; or used to make a dough smooth and even in consistency.

LARDING

A process of inserting match-like strips of salt pork about one-fourth inch in thickness into a dry meat or fish. These strips are called lardons, and are inserted either by making an incision in the surface and laying the lardon in the slashing or by the use of a larding-needle.

The pork is clamped into one end of the needle and is threaded into the meat, as in any sewing process.

MENU PLANNING

Menu planning doesn't have be complicated. A small investment of time can reap great rewards:

- *A menu plan saves money*: Reducing trips to the supermarket, a menu plan reduces impulse spending. Using leftovers efficiently cuts food waste, while planned buying in bulk makes it easy to stockpile freezer meals at reduced prices.
- *A menu plan saves time*: No dash to the neighbors for a missing ingredient, no frantic searches through the freezer for something, anything to thaw for dinner.

- *A menu plan improves nutrition*: Without the daily dash to the supermarket, there's time to prepare side dishes and salads to complement the main dish, increasing the family's consumption of fruits and vegetables.

DARE TO DO IT

For too many of us, making a menu plan is something we intend to do... when we get around to it. Instead of seeing menu planning as an activity that adds to our quality of life, we dread sitting down to decide next Thursday's dinner. "I'll do that next week, when I'm more organized."

Wrong! Menu planning is the first line of defense in the fight to an organized kitchen, not the cherry on the icing on the cake.

Take the vow: "I, [state your name], hereby promise not to visit the supermarket again until I've made a menu plan!"

START SMALL AND SIMPLE

Still muttering, "But I don't wanna..."? Break into menu planning easily by starting small and simple.

Think, "next week." Seven little dinners, one trip to the supermarket. Sure, it's fun to think about indexing your recipe collection, entering the data in a relational database and crunching menus till the year 2010, but resist the urge. Slow and steady builds menu planning skills and shows you the benefits of the exercise. Elaborate hoo-rah becomes just another failed exercise in home management overkill.

Where to start? The food flyers from your local newspaper. Try to make your menu plan and shopping list the day the food ads appear. You'll use the ads to get a feel for the week's sales and bargains. Use that feeling to guide your menu plan. This week in Eastern Washington, for instance, two local chain supermarkets are offering whole fryers for the low, low price of 59 cents a pound. Clearly, this is the week for Ginger Chicken and Fajitas, not a time to dream about Beef Stew and Grilled Pork Tenderloins.

MENU PLANNING BASICS

Okay, it's food ad day. Ready? Time to rough out a simple menu plan. The goal is two-fold: shop efficiently to obtain food required for seven dinner meals, while minimizing expenditure, cooking, shopping and cleaning time.

Here's the overview of the process:

- Scan the food ads for specials and sales. Rough out a draft menu plan: seven dinner entrees that can be made from weekly specials, side dishes and salads.
- Wander to pantry and refrigerator to check for any of last week's purchases that are languishing beneath wilting lettuce or hardening

tortillas. Check for draft recipe ingredients. Review your shopping list and note needed items.

- Ready, set, shop—but shop with an open mind. That 59-cent fryer won't look like such a bargain next to a marked-down mega-pack of boneless chicken breasts at 89 cents a pound. Be ready to substitute if you find a great deal.
- Return from shopping. As you put away groceries, flesh out the menu plan. Match it up with the family's calendar, saving the oven roast for a lazy Sunday afternoon, the quick-fix pizza for soccer night.
- Post the menu plan on the refrigerator door. Refer to it during the coming week as you prepare meals.

That's it! The bare bones of menu planning. You've made a draft plan, shopped from a list, retained flexibility in the marketplace, firmed up your plan and held yourself accountable.

COAST IN THE CALM OF A ROUTINE

Yes, there are some well-organized souls among us who don't make formal meal plans. Look close, and you'll discover that household meal service dances to a routine.

Sunday's a big dinner, and Tuesday gets the leftovers. Monday is burger night, and Wednesday sees spaghetti, year in and year out. Thursday's the day for a casserole, and Dad grills on Friday. Saturday night, it's take-out or pizza. Create a routine around your menu planning. Sure, you can try new recipes—just don't let your enthusiasm for the glossy pages of the cookbook con you into doing so more than twice a month. Cooking tried-and-true speeds dinner preparation and streamlines menu planning.

To do it, look for cues in the family schedule. At-home days with more free time can handle a fancy meal—or can signal soup, sandwiches and Cook's Night Off. Running the evening kid carpool is a great time to plan for pick-up sandwiches. Make the routine yours, and it will serve you well.

STAY FLEXIBLE

Menu plans aren't written in stone. So you're dodging cramps on the "big" cooking day? Swap it out with Pizza Night and go to bed early with a cup of herb tea. A posted menu plan promotes accountability, but family members will forgive you, as long as they get their postponed favorite a day or two later.

Build flexibility into your plan and serve the aims of thrift with Cook's Choice Night. Traditionally held the night before grocery shopping, you can slide a neglected dinner into Cook's Choice, or chop up the contents of the refrigerator for a clean-out stir-fry. Either way, you'll feel smug at your frugality and good planning.

MAKE IT A HABIT

Simple or not, a menu plan won't help you if you don't make one. Weekly menu planning is a good candidate for the Habit Patrol. Get into the habit of planning before you shop, and you'll get hooked—one addiction of great value.

RECYCLE MENU PLANS

After you've made menu plans for a few weeks, the beauty of the activity shines through: recycle them! Your family won't mind, and you'll save even more time and energy. Instead of an ambitious plan for 30-day menus, tuck completed menu plans in a file folder or envelope. Next time fryers are 59 cents a pound at the market, pull out the plan you made this week. Done!

BENEFITS

- You can extract a grocery list from your menu choices.
- Because you buy only what you need, less food is wasted.
- You know with plenty of notice what's for dinner – no more frantic 5 p.m. crazyness.
- Cooking is more enjoyable, because odds are you'll be more prepared.
- You'll have more variety, because you've planned it.
- It's healthier, because it cuts down on drive-thru runs.
- It's cheaper, too, because you're eating out less, you can menu plan around your coupons, and you can intentionally cook with seasonal ingredients.

MENU PLANNING TIPS

Use the menu planner and grocery list shown below as you follow these steps:

- Get out your cookbooks and plan several main meals. Make a list of these menu ideas on a menu planner. Plan some quick meals for busy nights. You also can double some recipes that freeze well, and save half for other busy nights when you don't have time to cook.
- Check your pantry for all the ingredients called for in recipes. Write down all the ingredients you need to purchase. If you notice you are running low on certain basic items, put these on your grocery list as well.
- Now fill in foods that you use to make breakfast, lunches, and snacks, such as eggs, cereals, breads, tuna, milk, and juice. List plenty of fruits and vegetables.
- Post this list on the refrigerator and add to it as you run out of foods or think of other things you need.
- Take the list to the store and stick to it. You save money by not making impulse buys. However, you don't need to be so rigid that you pass up a good sale item.

- When you get home, cut out and post the menu planner in your kitchen. You may want to write down page numbers from recipe books for quick reference. This way, whoever gets home first can start dinner.

The first few times you do this, it will seem like a fair amount of work. But the rewards are worth it, and you will become faster at the planning process. You can even save some of the menus and grocery lists and use them again in a few weeks.

FOOD SAFETY

Food safety is a scientific discipline describing handling, preparation, and storage of food in ways that prevent foodborne illness. This includes a number of routines that should be followed to avoid potentially severe health hazards. Food can transmit disease from person to person as well as serve as a growth medium for bacteria that can cause food poisoning. Debates on genetic food safety include such issues as impact of genetically modified food on health of further generations and genetic pollution of environment, which can destroy natural biological diversity. In developed countries there are intricate standards for food preparation, whereas in lesser developed countries the main issue is simply the availability of adequate safe water, which is usually a critical item.

REGULATORY AGENCIES

European Union

The parliament of the European Union (EU) makes legislation in the form of directives and regulations, many of which are mandatory for member states and which therefore must be incorporated into individual countries' national legislation. As a very large organisation that exists to remove barriers to trade between member states, and into which individual member states have only a proportional influence, the outcome is often seen as an excessively bureaucratic 'one size fits all' approach. However, in relation to food safety the tendency to err on the side of maximum protection for the consumer may be seen as a positive benefit. The EU parliament is informed on food safety matters by the European Food Safety Authority. Individual member states may also have other legislation and controls in respect of food safety, provided that they do not prevent trade with other states, and can differ considerably in their internal structures and approaches to the regulatory control of food safety.

United States

Federal Level Regulation

The Food and Drug Administration publishes the Food Code, a model set of guidelines and procedures that assists food control jurisdictions by providing a scientifically sound technical and legal basis for regulating the retail and food service industries, including restaurants, grocery stores and institutional foodservice providers such as nursing homes. Regulatory agencies at all levels of government in the United States use the FDA Food Code to develop or update food safety rules in their jurisdictions that are consistent with national food regulatory policy. The FDA, 48 of 56 states and territories, representing 79% of the U.S. population, have adopted food codes patterned after one of the five versions of the Food Code, beginning with the 1993 edition.

In the United States, federal regulations governing food safety are fragmented and complicated, just as to a February 2007 report from the Government Accountability Office. There are 15 agencies sharing oversight responsibilities in the food safety system, although the two primary agencies are the U.S. Department of Agriculture (USDA) Food Safety and Inspection Service (FSIS), which is responsible for the safety of meat, poultry, and processed egg products, and the Food and Drug Administration (FDA), which is responsible for virtually all other foods.

The Food Safety and Inspection Service has approximately 7,800 inspection program personnel working in nearly 6,200 federally inspected meat, poultry and processed egg establishments. FSIS is charged with administering and enforcing the Federal Meat Inspection Act, the Poultry Products Inspection Act, the Egg Products Inspection Act, portions of the Agricultural Marketing Act, the Humane Slaughter Act, and the regulations that implement these laws. FSIS inspection program personnel inspect every animal before slaughter, and each carcass after slaughter to ensure public health requirements are met. In fiscal year (FY) 2008, this included about 50 billion pounds of livestock carcasses, about 59 billion pounds of poultry carcasses, and about 4.3 billion pounds of processed egg products. At U.S. borders, they also inspected 3.3 billion pounds of imported meat and poultry products.

State and Local Regulation

A number of U.S. states have their own meat inspection programs that substitute for USDA inspection for meats that are sold only in-state. Certain state programs have been criticized for undue leniency to bad practices.

However, other state food safety programs supplement, rather than replace, Federal inspections, generally with the goal of increasing consumer confidence in the state's produce. For example, state health departments have a role in investigating outbreaks of food-borne disease bacteria, as in the case of the 2006 outbreak of *Escherichia coli* O157:H7 from processed spinach. Health departments also promote better food processing practices to eliminate these threats.

In addition to the US Food and Drug Administration, several states that are major producers of fresh fruits and vegetables have their own state programs to test produce for pesticide residues. Restaurants and other retail food establishments fall under state law and are regulated by state or local health departments. Typically these regulations require official inspections of specific design features, best food-handling practices, and certification of food handlers. In some places a letter grade or numerical score must be prominently posted following each inspection. In some localities inspection deficiencies and remedial action are posted on the Internet.

China

Food safety is a growing concern in Chinese agriculture. The Chinese government oversees agricultural production as well as the manufacture of food packaging, containers, chemical additives, drug production, and business regulation. In recent years, the Chinese government attempted to consolidate food regulation with the creation of the State Food and Drug Administration of China in 2003, and officials have also been under increasing public and international pressure to solve food safety problems. However, it appears that regulations are not well known by the trade. Labels used for "green" food, "organic" food and "pollution-free" food are not well recognized by traders and many are unclear about their meaning. A survey by the World Bank found that supermarket managers had difficulty in obtaining produce that met safety requirements and found that a high percentage of produce did not comply with established standards.

Traditional marketing systems, whether in China or the rest of Asia, presently provide little motivation or incentive for individual farmers to make improvements to either quality or safety as their produce tends to get grouped together with standard products as it progresses through the marketing channel. Direct linkages between farmer groups and traders or ultimate buyers, such as supermarkets, can help avoid this problem.

Governments need to improve the condition of many markets through upgrading management and reinvesting market fees in physical infrastructure. Wholesale markets need to investigate the feasibility of developing separate sections to handle fruits and vegetables that meet defined safety and quality standards.

Australia

Australian Food Authority is working toward ensuring that all food businesses implement food safety systems to ensure food is safe to consume in a bid to halt the increasing incidence of food poisoning, this includes basic food safety training for at least one person in each business. Smart business operators know that basic food safety training improves the bottom line, staff take more pride in their work; there is less waste; and customers can have more confidence in the food they consume. Food Safety training in units of competence from a relevant training package, must be delivered by a Registered Training Organization (RTO) to enable staff to be issued with a nationally-recognised unit of competency code on their certificate. Generally this training can be completed in less than one day. Training options are available to suit the needs of everyone. Training may be carried out in-house for a group, in a public class, via correspondence or online.

Basic Food Safety Training includes:

- Understanding the hazards associated with the main types of food and the conditions to prevent the growth of bacteria which can cause food poisoning
- The problems associated with product packaging such as leaks in vacuum packs, damage to packaging or pest infestation, as well as problems and diseases spread by pests.
- Safe Food handling. This includes safe procedures for each process such as receiving, re-packing, food storage, preparation and cooking, cooling and re-heating, displaying products, handling products when serving customers, packaging, cleaning and sanitizing, pest control, transport and delivery. Also the causes of cross contamination.
- Catering for customers who are particularly at risk of food-borne illness, including allergies and intolerance.
- Correct cleaning and sanitizing procedures, cleaning products and their correct use, and the storage of cleaning items such as brushes, mops and cloths.
- Personal hygiene, hand washing, illness, and protective clothing.

People responsible for serving unsafe food can be liable for heavy fines under this new leglislation, consumers are pleased that industry will be forced to take food safety seriously.

CONSUMER LABELING

United Kingdom

Food stuffs in the UK have one of two labels to indicate the nature of the deterioration of the product and any subsequent health issues:

Best before indicates a future date beyond which the food product *may* lose quality in terms of taste or texture amongst others, but does not imply any serious health problems if food is consumed beyond this date.

Use by indicates a legal date beyond which it is not permissible to sell a food product due to the potential serious nature of consumption of pathogens. Leeway is sometimes provided by producers in stating display until dates so that products are not at their limit of safe consumption on the actual date stated. This allows for the variability in production, storage and display methods.

United States

With the exception of infant formula and baby foods which must be withdrawn by their expiration date, Federal law does not require expiration dates. For all other foods, except dairy products in some states, freshness dating is strictly voluntary on the part of manufacturers.

In response to consumer demand, perishable foods are typically labeled with a Sell by date. It is up to the consumer to decide how long after the Sell by date a package is usable. Other common dating statements are Best if used by, Use-by date, Expiration date, Guaranteed fresh <date>, and Pack date.

ISSUES ASSOCIATED WITH SELL BY/ USE BY DATES

The UK's Waste & Resources Action Programme, 33% percent of all food produced is wasted along the chill chain or at the consumer. At the same time, a large number of people get sick every year due to spoiled food. WHO and CDC in the USA alone there are every year 76 million of food bourne illnesses leading to 325'000 hopitalizations and 5000 deaths

UK Government to Replace Sell by/ use by Dates?

UK minister Hilary Benn the use by date and sell by dates are old technologies that are outdated and should be replaced by other solutions or disposed of altogether.

How to Enhance Food Safety

There is a number of ways to enhance sell by and use by dates. These include better education of consumers on how to use, transport, and store fresh food products, but also by enhancing the use by and sell by dates by adding to the package smart indicators such as TTIs (Time Temperature Indicators). These show through a visible color change whether the product is still fresh. TTIs are already in use by retailers and food producers in France, Switzerland, and other countries in western Europe.

EVALUATING SUPPLIERS

Supplier evaluation is not an easy process. Many different formulas and techniques can be used. One effective method assigns suppliers to four basic categories based upon their level of performance in key areas, such as delivery, quality and responsiveness.

A supplier is labeled a "full partner" if it meets all expectations. An "associate partner" is a firm that needs a little work to bring it up to full partner status. Additional categories include "high risk" and "incapable" suppliers. A high risk source might continue to be utilized for current production only, based upon an internal risk assessment. But, you might not want to award this supplier any future business.

An incapable supplier should be dropped as soon as possible. The first step is to agree upon a set of criteria for each of these categories. The following characteristics should be considered:

Full Partner:

- Goals are compatible with your organization's future plans.
- Demonstrates 100 percent on-time delivery performance and is capable of participating in your organization's automatic ordering systems.
- Fully ISO 9000 or QS 9000 approved.
- Provides full field response and support within 24 hours of notification.
- Provides assistance with new designs.
- Provides proto-samples within one week or less (depending upon tooling).
- Consistently meets the full partner criteria in five key areas: technology, quality, responsiveness, delivery and lowest total cost.

Associate Partner:

- Satisfies most of the full partner criteria as agreed upon by management.
- Supplier's management and its employees demonstrate an ongoing commitment to continuous improvement in quality and delivery. Develops goals, objectives and action plans to meet or exceed the goals and objectives of your organization.
- Supplier's services or products fulfill a strategic need in your organization.
- Requires little effort and expense to achieve full partner status.

High Risk Supplier:

- Supplier's goals and action plans are not compatible with your organization.
- No benefit in developing these sources to associate partner or full partner status.
- Current quality and delivery are acceptable to sustain present production.
- Not economically feasible to move tooling for parts these sources are currently producing.

Incapable Supplier:

- Supplier does not meet quality and delivery goals.
- Does not demonstrate a desire or a capability to improve deficient service areas, such as quality and delivery.
- Has no interest in obtaining ISO 9000 or QS 9000.
- This supplier should be dropped immediately.

Once you have agreed upon these categories and established criteria, you're ready to focus on the topic that many accountants claim is the most important—cost. However, cost must be taken to the next step.

In other words, you need to look at the total cost—not only piece price. The lowest total cost of a product includes not only its initial procurement cost, but all those factors that went into getting it to market and keeping it there.

Therefore, you must take into consideration these added factors. Lowest total cost can be calculated as: LTC = Piece Price x IV. In this equation, IV is a composite index determined by tables developed and agreed upon within your organization.

For example, if you select quality, delivery and proto lead time as the factors you want to assign an IV to, then the formula would appear as: LTC = Piece Price × Quality IV × Delivery IV × Proto Lead Time IV.

If one supplier exhibits a 315 PPM performance level, a 100 percent on-time delivery record and a 10-day lead time, its composite index would be 1.04. If a second supplier exhibits a 12,400-PPM performance level, a 20 percent on-time delivery and a 20-day lead-time, its composite index would be 3.65.

If the first supplier's part cost $0.06 each and the second supplier's part cost $0.03 each, whose part is better? By following this formula, you can develop tabulated charts that track lowest total cost for all your suppliers. This can be a very effective evaluation tool.

FULL PARTNER CRITERIA

Suppliers qualify as "full partners" if they meet the following key criteria.

Technology:

- Demonstrates an ongoing ability to assimilate state-of-the-art technology into their business.
- Demonstrates an ability to quickly develop new processes and modify existing processes as necessary to create a sustained competitive advantage for customer and self.
- Strong commitment to R and D.
- Demonstrates a willingness to invest in the business and grow.
- Willingness to pursue mutual product and process development.
- Complete understanding of process capabilities.
- Provides documented design guidelines usable by customer engineers.
- Full ISO 9000 or QS 9000 approved.

Quality:

- Consistently meets or exceeds your organization's quality goals.
- Knows whether or not the parts being produced for your organization meet or exceed the letter and intent of the specification.

- Does not ship out-of-spec parts.
- Has well-documented processes (both manufacturing and administrative).
- Demonstrates a willingness to say "no" and present alternatives to new designs that lack process capability.

Responsiveness:

- Demonstrates an ability to interact directly with your engineers and maintain ongoing business relationships.
- Actively supports your remote manufacturing sites.
- Provides timely responses to technical and business issues.
- Demonstrates a high level of commitment to your organization.
- Is a long-term thinker and is financially stable.
- Demonstrates flexibility to quick changes.

Delivery:

- Willingness to work with your organization to optimize lead times.
- Consistently meets delivery goals.
- Allows order flexibility within acceptable limits.
- Has business systems in place to accurately schedule product and procure material in an economical manner.
- Notifies your organization in advance of shipments delayed for any reason.

Lowest Total Cost:

- Demonstrates a commitment to continuous cost reductions.
- Willing to work with you to perform value engineering for new and existing designs.
- Openly shares cost structures and pricing models.

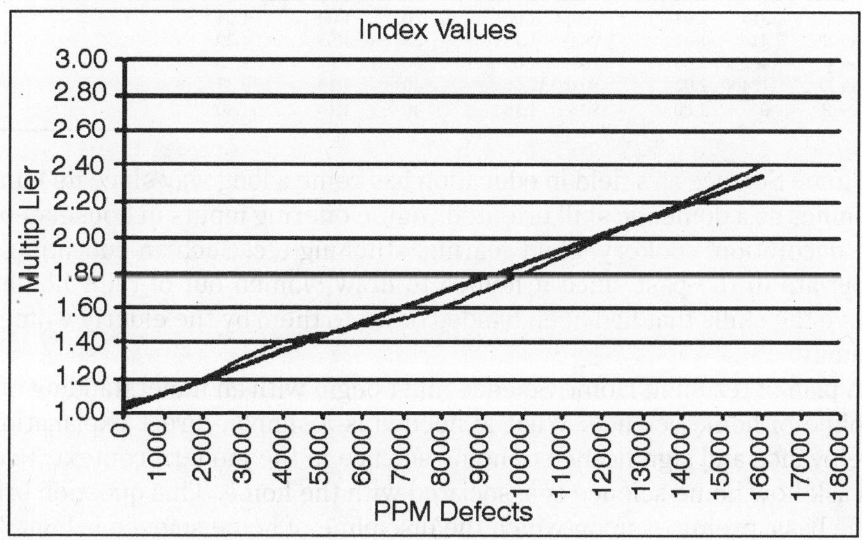

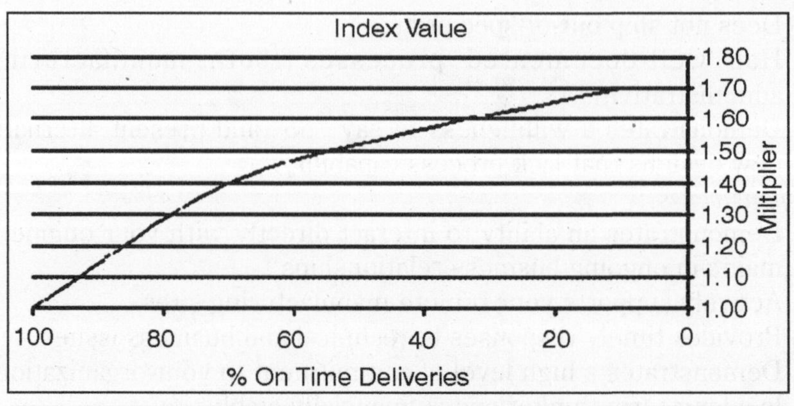

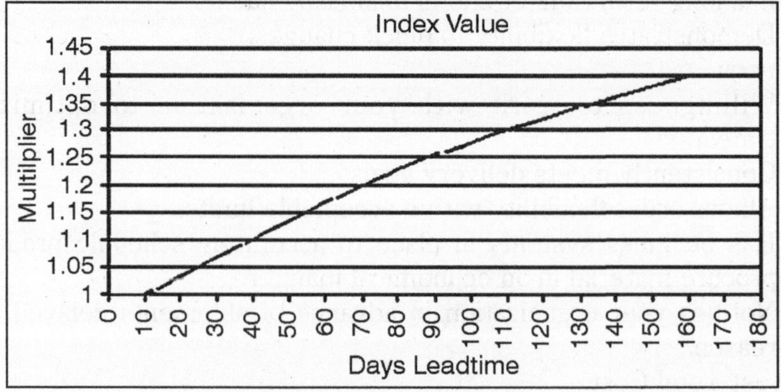

Table. Supplier Cost Index

Supplier Name	PPM	Multiplier	% on Time	Multiplier Lead Time	Days	Multiplier	Composite Index
Supplier A	315	1.04	100.0	1.00	20	1.00	1.04
Supplier B	41	1.01	96.0	1.13	28	1.05	1.20
Supplier C	497	1.06	100.0	1.00	38	1.07	1.13
Supplier D	12400	2.02	50.0	1.50	40	1.10	3.33
Supplier E	0	1.00	90.9	1.18	40	1.1	1.30

Home Science as a field in education has come a long way since its humble beginnings as a domestic skill oriented course offering inputs in housekeeping, home decoration, cookery, child rearing, stitching etc. Such an education was appropriate in the past since it helped to draw women out of their homes to improve the skills that had been handed down to them by the elderly women in the family.

A plan of teaching Home Science must begin with an understanding of the discipline of home science. What is needed is a simple, direct explanation of the relevance and significance of home science in the modern context. People often ask how home science is associated with the home. This question brings out the basic premises upon which the discipline of home science is built. The science of the home is concerned with the maintenance and enrichment of human relationships through the development and judicious use of all available human and material resources to achieve a maximal satisfying life for all members of the family.

Home science education prepares youth for the greatest of all vocation - Home Making. It orients young girls and boys towards preparation for several professions- teaching, nursing, dietetics, research, welfare, management, art application, extension work and communication. There are several ways of managing homes. Men as well as women have played the roles of breadwinners and homemakers.

It has become necessary for men to share housework, to enable women as professional persons to contribute to the nation. Traditionally, the field of Home Science encompasses five areas, namely, Food and Nutrition, Human Development and Family Studies, Fabric and Apparel, Resource Management and Communication and Extension.

All these domains have their specific content and focus that contribute to the study of the individual and the family in Indian sociocultural context.

To enable students to develop a holistic understanding of life in the home and outside the boundaries between different areas of the discipline need to be dissolved.

A special effort should be made to communicate respect for every students life at home and in society by making the curriculum appropriate for both boys and girls, living in different contexts, including those who are homeless.

MEANING

Home science, or the science of a home in simple words, includes all the things that concern yourself, your home, your family members and your resources. It aims at getting maximum satisfaction for you and your family members through the efficient and scientific use of your resources. This combination of art and science holds true in every sphere of your life like any of the following:-

- The house that you live in;
- The food that you eat;
- The clothes that you wear;
- The family that you care for;
- The resources that you use;
- The skills that can lead to a successful career.

IMPORTANCE OF HOME SCIENCE

Home Science gives knowledge for facing new challenges, to cope with knowledge explosion, technological advancements, new developments and growing needs of individuals for successful living in society.

Importance for individuals

Home Science will give you scientific knowledge and develop skill for efficient performance of household responsibilities. Skill is the process of applying theory based knowledge into your daily life. For example, countering social pressures to adopt unhealthy eating practices by refusal is a skill. Seeking services for help with sexual issues is another skill.

Importance for home and family

The emphasis in Home Science is to strengthen home and family life by making an optimal use of available resources. It will help you to apply knowledge of different sciences for improving home and family environment, health, growth and development of individuals and in managing your household resources. This is the only subject that deals with food, clothing, shelter, health, human relationships, household resources and concerns of individuals that exist within a home and in a family

Importance for economic stability

Different areas of Home Science prepare you for a variety of jobs. The economic stability of the family can be ensured by undertaking a job or self employment. This will further lead to raising the living standard of the family and quality of life.

MISCONCEPTIONS

Although home science has evolved into a very vast and detailed area of study, the layperson still has many misconceptions about the subjects. It is because is not able to achieve the status and respects that it correctly deserves.

FACT

Cooking, laundry and needle work are only a very small part of what is taught in home science. In fact, we learn about everything that touches our life-

- The basics of how our body works
- What we need to eat to remain healthy
- Factors which our selection of fabrics for various uses
- Concerned about raising children safely and sensibly
- Learn to manage our resources like time, energy & money

FACT- Although students do learn many things by observing their mother and grandmother at work, done traditionally which have a sound scientific basis.

For example- Pickles are covered with a layer of oil. When asked why, mother may not be able to explain. But home science will explain that the layer of oil prevents contact of pickle with air. This prevents spoilage.

RELEVANCE OF HOME SCIENCE

Traditionally Indian society had clearly demarcated areas of work for both men and women. However, with the changing social and economic environment, managing a home requires the efforts of all the family members. Today, more and more women are working- whether they go to an office or work from home. Therefore, gender based stereotype roles do not hold relevance any longer. Today, you can learn to manage your own resources better by studying various areas of Home Science. If you face any problem, Home Science gives you the right tools to solve. In doing so, you become more organized and efficient. Home science opens up a large arena of possible job opportunities for both men and women. This knowledge helps in the improvement of not only your home and personal life but also your professional life.

Home Science is for both Boys and Girls:

- Both have to succeed in an increasingly competitive world
- Both have to share household responsibilities and tasks
- Both need to manage resources
- In case of a problem they have to solve it together with other family members.
- Home Science offers varied vocational and professional avenues for both boys and girls
- Joint knowledge and skills help to improve the standard of living.

You and your family members can achieve personal satisfaction and use the skills gained through Home Science to improve your family's economic condition as well. As in Home Science you will learn several skills that can enable you to set up home based enterprises or consultancies like a boutique, catering unit, crèche, day care centre, etc.

HOME SCIENCE- A DISCIPLINE AND PROFESSION

- Home Science covers a few areas of specialization such as Food and nutrition, Communication and Extension, Resource Management, Human Development, Fabric and apparel science. We learn and develop good interpersonal relationships within and outside the family. We also learn to manage our resources like time, energy and money so that we get maximum satisfaction. Home science also involves learning the skills intending all this useful to our family or community, especially the underprivileged for better life.
- We value our traditional knowledge, experiences offered to us by our parents and grandparents. We should continue and value this. The new advance in science and technology and management equip us

better and deal with the changing social and economic environment and meet the challenges of the 21st centaury.

- Home science syllabus draws its strength from both science and arts discipline. This enables the student to develop the ability and understand the concept as well as apply them in various contents situation. This gives home science students an edge above all other disciplines. This prepares them for vast range of opportunities unlike other discipline.

AREAS OF SPECIALIZATION IN HOME SCIENCE

By now you must have understood that Home Science is an important area of study. Let us now find out in detail the specific areas that you have to study in this subject. There are five major components or areas of specialization in Home Science:

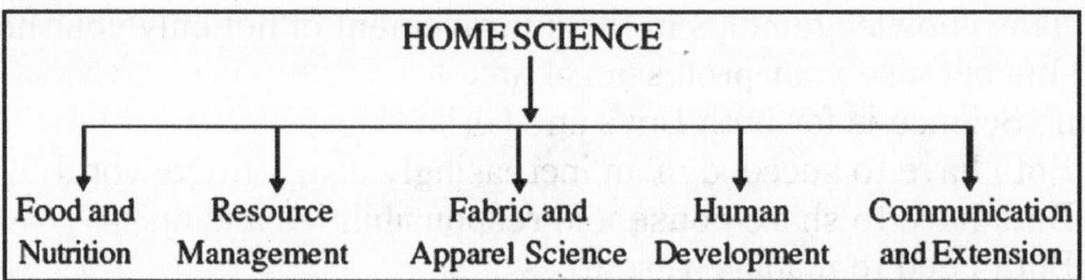

Today, Home Science is so advanced that each specialization is a domain in its own with its areas of specialization and vocation/professional opportunities. Some of the areas of further specialization are as follows:

Main Branch	*Area of further specialization*
Food and Nutrition	• Food Science • Nutrition- Clinical Nutrition and Community Nutrition • Institutional food service
Fabric and Apparel Science	• Clothing Construction • Textile Science • Textile Designing • Garment Designing • Care and Maintenance of Clothes
Resource Management	• Resource Management • Housing and Equipment • Interior Decoration • Consumer Education
Human Development	• Child - Welfare • Adolescence, Marriage and Family Guidance • Care of the Elderly • Care of special children
Communication and Extension	• Media for communication • Programme planning and evaluation • Training and capacity building • Management of community service organisations

CAREER OPPORTUNITIES IN HOME SCIENCE

Before you study the different opportunities available in the subject of Home Science, you must understand the concepts of wage employment and self-employment. To understand these concepts, let us take an example - a tailor who sets up a tailoring shop is self-employed whereas he/she is said to be in wage employment when he/she works in a garment factory and gets a salary or wage. Wage employment means that you work for another person and receive wages or salary for your services. Self-employment means that you are the owner of an enterprise which you run and finance. You have already read that specializing in Home Science may choose to work in salaried jobs or as self-employed entrepreneurs or as consultants using any of the skills that they may have developed.

FOOD AND ITS NUTRIENTS

FOOD AND ITS IMPORTANCE FOR HEALTHY LIVING

Food satisfies our hunger and provides energy for day to day living. It is also an important part of all our social occasions. When you go to a friend's house, you are generally greeted and offered something to eat and drink. So food serves a social function. Similarly when you are away from your home you not only miss your family but also food. Thus, food is important in our life. Can you elaborate the importance of food in our life? Yes, you are right. We eat to satisfy our hunger, feel happy and even celebrate special occasions. Let us now learn about the various functions of food.

- Food gives us energy to work: We need energy to do our work. We need energy for walking, playing, eating, working in thehouse or outside and for otheractivities.
- Food helps in the growth and repair of tissues: A small child grows into an adult. Do you know the relationship between the growth of our body and the food we eat? The body is made up of thousands of small cells. New cells are added to help the body to grow in size. We need food for the formation of new cells. In addition to growth, new cells are made to replace the dead and damaged cells. Therefore, food helps not only in growth but also in the repair of tissues.
- Food gives strength to fight against diseases: We are always surrounded by disease causing organisms. The food we eat helps us in providing protection against them. If we fall ill, food helps us to recover from illness.
- Food helps the body to function normally: Do you think we need energy even when we are at rest? Yes you are right, it is because the vital organs inside your body are always working. For example, our heart is pumping blood, stomach is digesting food and lungs are

breathing in air. These organs need energy to perform their role which is provided by the food we eat. and stopped communicating with others. Finally they had to be admitted in a hospital in a state of acute depression and nutritional deficiencies. Anuradha died due to severe under nutriton. Shonali survived and had to be treated by doctors, nurses and social workers for a long time for both depression and malnutrition. She had to learn to eat proper meals all over again. Think about this situation and try to find answers on the basis of the functions performed by food:

Q. What could be the reasons due to which the sisters stopped eating?

Q. Do you stop eating food when you are angry or sad?

You have seen that food performs many functions in our body and all of them are important for not only our existence but also for a healthy life.

- Food satisfies hunger: We eat food to satisfy our hunger. It also provides mental and psychological satisfaction. Therefore is the prime need to be satisfied in life.
- Social functions: Food has always been a central part of our social existence. It helps to relax and creates a friendly mood. People take special care in planning and cooking food for social occasions, so food promotes a sense of community.
- Psychological functions: In addition to social functions food must also satisfy our items emotional needs. These include a sense of security, love and attention. Familiar food items make us feel secure. A baby gets emotional security while being breast feed. Anticipating food needs and fulfilling them are expressions of love, care and attention. Similarly, sharing of food is a token of friendship and acceptance.

NUTRITION AND NUTRIENTS

Have you ever thought what happens to the food we eat? How do we get energy from the food? Now we will discuss about the science behind food and its functions. We can define nutrition as the science which tells us about what happens to the food when it enters our body and thereafter. How is food digested and utilized in our body? How does one grow from a little child to a young adult and then an old person? The food we eat contains many chemical substances. These chemical substances are known as nutrients. We can say that nutrients are invisible compounds in the food which are necessary for keeping the body healthy. These nutrients have different names and functions to perform in our body. The common nutrients in our food are:

- Carbohydrates
- Proteins
- Fats

- Vitamins
- Minerals
- Fibre
- Water

FUNCTIONS AND SOURCES OF NUTRIENTS

We will now learn about the functions of various nutrients and their sources.

Carbohydrates

Carbohydrates are as important to our body as fuel is to a car. As fuel makes a car run, similarly carbohydrates provide energy to our body, which keeps it going throughout the day. Which food items in our diet contain carbohydrates? Yes, foods like potatoes, rice and sweets give carbohydrates. There are two types of carbohydrates in food sugar and starch. Sugar: Sugar is also called simple carbohydrate. Fruits, honey and sugar are some sources of sugar. Starch: starch is called complex sugar. Plants store energy in the form of starch. When we eat plant products containing starch, our digestive system breaks down starch into glucose. This glucose is absorbed in blood and provides energy. Some foods give carbohydrates to our body in the form of starch. These are cereals, potatoes, sweet potatoes etc. Some other foods give carbohydrates in the form of sugar. For example sugar, honey, jam, jaggery, etc. One gram of carbohydrate gives four kcal. An adult person needs 400-420 gms of carbohydrates everyday.

Proteins

Proteins are needed for muscle building and repairing worn out cells and tissues. Our muscles, organs and even blood are made up of mostly proteins. If we do not eat proteins our body will not be able to repair damaged cells or build new ones. Proteins in our diet come from both animal and plant sources. The proteins from animal source are known as animal proteins and the proteins from plant sources are known as plant proteins. One gram of protein gives four kilo calories of energy (kcal). An adult needs 1gm of protein for every kilogram of body weight per day. For example if Shivali weighs 52 kg then her protein requirement would be approximately 52gms per day.

Fat

One gram of fat gives you nine kcal. Fat which we get from animals are called animal fat. Fat which we get from plants are called plant fat. Although fat gives more energy than carbohydrates, you eat very little fat and hence fat is not the major source of energy in our body. Food containing large amount of fat can lead to obesity. An adult needs only 30 gms (2 table spoons) of fat everyday. Do you know that chocolate, biscuits, pakoras and patties also contain fat?

Minerals

Minerals are present in all body tissues and fluids. Minerals like calcium and phosphorus are found in bones and teeth. Iron is a mineral which is found in blood as a part of the red pigment haemoglobin. Minerals occur in food as salts.

Calcium and Phosphorus are present in the body in large amounts especially in bones and teeth and are known as macro or major minerals. Iron and Iodine are found in the body in smaller quantities, hence are known as micro minerals. Ten year old Ajay and his eight year old sister Priyanka asked their mother why she insists that they should consume more milk and green vegetables. Mother showed them their childhood photographs. The children were surprised to see the photographs. They asked their mother how had they become so tall and big! Their mother explained that it's the milk and green vegetables that helped them grow so well.

CALCIUM

We all require calcium in our daily diet as it performs the following functions in our body:

Functions:

- Helps bones to grow and become strong.
- Makes teeth healthy and strong.
- Helps in clotting of blood. You may have noticed that in case of a minor injury blood stops flowing in few seconds and a hard crust is formed on the wound this is known as clotting of blood, for which calcium is needed. Blood clotting can prove to be a life protecting function.
- It also helps in the movement of muscles.

Calcium is required in large quantity for optimum growth during childhood. A pregnant woman also needs additional calcium to support the growth of foetus. Absorption of calcium is reduced as a person grows old. So, we should be careful regarding the calcium needs of the elderly persons.

Sources:

- Milk and its products like paneer, cheese, curd, lassi, chhach, butter milk etc. These are the best sources of calcium.
- Green leafy vegetables like spinach, fenugreek, curry leaves and coriander leaves are also good sources of calcium.

IRON

Functions:

- Iron is a very important mineral. It is needed for the formation of a compound called haemoglobin in our blood which helps in carrying oxygen to all parts of the body. Do you know in adolescence (10-19 years) iron requirement of the body increases? The need for iron increases specifically for girls. We must include iron rich foods in our daily diet.

- Green leafy vegetables like spinach, mustard leaves, fenugreek, mint, etc.
- Whole cereals and legumes
- Liver, heart, kidney and egg yolk.
- Gur/jaggery
- Dates and pomegranate.

IODINE

Iodine is required for normal functioning of our brain and the growth of our body. Deficiency of iodine leads to a disease called cretinism.

Functions:

- Iodine is needed for proper functioning of thyroid gland in our body.
- Sea foods like fish and sea weeds.
- Plants which grow in iodine rich soil.
- Salt fortified with Iodine.

VITAMINS

These are the substances which are required in very small amounts in our diet, but are essentials for proper functioning of the body. Vitamins cannot be produced by the body, therefore must be present in our diet. Vitamins are classified into two groups on the basis of their solubility in fats and water.

- Fat soluble vitamins are those vitamins which are soluble in fat e.g. Vitamin A, D, E, and K.
- Water soluble vitamins are those vitamins which are soluble in water e.g. Bcomplex vitamins and vitamin C.

You must know the functions and sources of vitamins. Table 2.1 lists vitamins, their functions and sources.

DIETARY FIBRE

Shruti's grandmother is an active and cheerful old lady. However, she was uncomfortable as she had not passed stool since last three days. She realized that the in last two-three days she had not been taking enough fresh fruits, vegetables and water. Fresh fruits and vegetables contain fibre that helps in relieving constipation Dietary fibre, also known as roughage, is the indigestible carbohydrate present in food. Fibre is present in foods of plant origin only. Foods rich in fibre are:

- Whole wheat flour, dalia (porridge)
- Fruits such as guava, apples, pineapples,banana
- Whole pulses like lentil, rajmah, bengal gram

- Vegetables such as peas, beans, carrots, cauliflower, green leafy vegetables

When foods are processed there is loss of fibre, vitamins and minerals. For example unpolished rice has higher fibre content than polished rice. Can you think of some other examples? Wheat flour is something many of us consumes everyday. Often people sieve the flour and then make chapaties out if it. In the process, wheat bran which is rich in fibre is removed. Thus, reducing the fibre content of the flour. The skin of fruits and vegetables is also rich in fibre. Fibre performs many essential functions in the body;

- It increases the bulk and softness of stool, thus making it easier to pass;
- Foods which contain fibre require more chewing and thus have a high degree of satiety.

High fibre diet helps in the prevention of:

- Constipation
- Cancer of the large intestine
- Diabetes
- Obesity

WATER

Mohan's mother was travelling to Jaipur in Rajasthan in the month of May for some work. She packed two bottles of water along with her lunch. Mohan asked his mother about the necessity of carrying the water bottles. Mother explained that during hot weather, our body sweats a lot. This results in depletion of body water. We should drink plenty of water to replenish the loss. Body water need can also be fulfilled by beverages like buttermilk (lassi), milk, fruit juices etc. On the other hand, our body's need for water is less during winters. Now Mohan understands that water requirement vary with the season. On an average, one should drink about 8-10 glasses of water everyday. Water is not a nutrient, yet it is very important for our body because;

- It is a constituent of all body fluids.
- It helps to digest food and takes the nutrients from the food to the different cells of your body.
- It helps to keep our body temperature constant. In summer, when we sweat, extra heat is removed from the body.
- It helps to remove the waste products from your body in the form of urine.

DEFICIENCY DISEASES

Geetika wants to lose weight and often skips her meals. Her teacher complained that she is not alert and attentive in the class. One day, she fainted during morning assembly. Her mother was called by the teacher. The mother explained that for the last four months, Geetika was eating very little. She had become very weak. The teacher advised Geetika that by reducing her food intake, she was not getting adequate nutrients essential for her body. This was causing weakness.

This would have detrimental effect on her physical, mental and emotional well being. Geetika became aware of the importance of a nutritious diet for her overall well being. She realized that each nutrient has a specific role in keeping us healthy. If you do not take these nutrients in the diet, do you know what will happen? Yes, you are right.

The body will not be able to function properly and you will fall sick. Interestingly, if you start eating the food with the missing nutrients, the symptoms of the disease start disappearing. This is the reason why such diseases are called deficiency diseases.

The deficiency is caused due to shortage or absence of certain nutrients in our daily diet. Deficiency disease is a condition which occurs in the body when a particular nutrient in our daily meals has been missing for a long period. Often the deficiency is reversible if the missing nutrient is detected early and missing nutrient is adequately replaced. Children and old people are more prone to deficiency diseases. Let us now study some common deficiency diseases.

Protein Energy Malnutrition (PEM)

When there is lack of proteins and energy in the diet for a long time a deficiency disease known as Protein Energy Malnutrition (PEM) occurs. This deficiency occurs mainly in children below 5 years of age. In our country PEM is widely prevalent in children belonging to poor and underprivileged communities. When the child is suffering from PEM, often diarrhea and other infections occur due to low body resistance to diseases. PEM is of two types-

- When the protein intake is deficient but calories or energy from carbohydrates are sufficient, the condition is known as Kwashiorkor. The child dvelops a pot like belly due to this deficiency diseases.
- When protein and energy both are insufficient for a prolonged period then a disease condition known as Marasmus occurs. In both these conditions there is growth failure, child does not grow to full potential.
- A child loses weight and becomes thin when there is deficiency of carbohydrates in the diet. Yet another kind of malnutrition is consumption of excessive calories. If there is excess of carbohydrates in the diet then the person becomes obese or fat. This is the condition mostly found in the well-to-do families. This is increasingly becoming a problem in urban areas where more and more people are leading inactive or more sedentary lines.

Vitamin A deficiency

Vitamin A deficiency is also found mainly in children. Pregnant women are also susceptible to this deficiency. Since Vitamin A is required for growth and development of the body.

Lack of this vitamin results in stunted growth. Deficiency of Vitamin A also leads to dryness of conjunctiva (the white portion of the eye). At a later stage the conjunctiva may become wrinkled. The cornea (black part of eye) may also become cloudy.

At this stage the problem of night blindness sets in, i.e., the child cannot see in dim light. Improvement in the condition can occur if a diet rich in Vitamin A is taken. You would be sad to know that more than 40,000 children in India become blind every year because of vitamin A deficiency. This could be easily prevented.

Iron deficiency anaemia

Iron deficiency anaemia is another major nutritional problem in our country. It is found in women and children in all income groups. Major causes of anemia in our country are:

- Inadequate intake of dietary iron or its poor absorption
- Hook worm infestation mainly found in young children and adults
- Excessive blood loss due to accidental injury, during child birth in women

You have already studied that iron is required for hemoglobin formation. In iron deficiency anemia, there is not enough haemoglobin in blood, therefore, the supply of oxygen to the cells is reduced. When this happens, we feel tired, restless and fatigued all the time. Therefore, we cannot work to our full potential.

Vitamin B Complex deficiency

Low intake of Vitamin B rich food in our diet is the main cause of this deficiency. Common symptoms of vitamin B complex deficiency are sore mouth, sore tongue, redness of the tongue and cuts at the angles of mouth. Some accompanying symptoms like diarrhea and dysentery may also occur. One does not feel hungry and has difficulty in digesting food. In earlier times a disease known as Beri Beri was quite prevalent in areas where polished rice was the staple food. This deficiency is no longer seen now. Consumption of brown rice instead of white rice prevents the occurrence of this disease.

Vitamin C deficiency

In the absence of fresh fruits and vegetables in our diet often deficiency of Vitamin C occurs. It is called Scurvy. Our gums, bones and teeth become weak. Gums swell up and start bleeding. Wounds also do not heal quickly. The only solution is to include fresh fruits and vegetables in our diet on regular basis.

Iodine Deficiency Disorders (IDD)

Iodine is very essential for the proper functioning of the body. Deficiency of iodine leads to a number of disorders like goitre, cretinism, mental retardation and deaf mutism. Goitre is generally seen among adolescents, young adults and children. More females than males are affected. A severe deficiency of iodine during prenatal life may lead to Cretinism (severe mental retardation). In India there are certain regions where goiter is quite prevalent areas. The foothills of Himalayas and the areas where there are frequent floods are the most goiter prone areas. Recent surveys have shown that areas in Andhra Pradesh, Madhya Pradesh, Maharashtra, Bihar, Gujarat and Kerala also have regions where goiter is prevalent. The only solution to this problem is to eat iodized salt. For this reason our Government has banned the used of non-iodized salt.

NEED FOR EATING VARIETY OF FOOD

Now you know the different nutrients and also the functions they perform in our body. You also know what happens when these nutrients are missing in our diet. So what should we do to remain healthy? We should eat food which provides all the nutrients. As you have seen, there is no single food which gives all the nutrients. You must therefore, eat variety of food items so that our body gets all the nutrients in the required amounts and we do not suffer from deficiency diseases.

RELATIONSHIP BETWEEN NUTRITION AND HEALTH

Knowledge of nutrition helps us to know the type and quantities of different foods to be taken in the diet in order to maintain good health. The World Health Organization (WHO) has defined health as "The state of complete physical, mental and social well being and not just the absence of disease". Hence, to be healthy in the real sense, one must not only eat right but also have a healthy mental and social outlook.

2

Family Education and Management

CONCEPT OF FAMILY LIFE

Family has been regarded as the cornerstone of society through several social thinkers. Marriage flanked by an adult man and an adult woman is the foundation of the formation a family from time immemorial. It is in the family that children are born, cared, reared, and provided education to life. In excess of the years global societies have witnessed shifts in family patterns. Radical changes are being brought into lifestyles through progress and reforms.

The recent phenomenon of globalization and consumerism has created diversity of family and household shapes. Beside with these changes social troubles and personal difficulties continue to augment and multiply. It would be unimaginable to earlier generations to understand the extent of changes taking place in the areas of family life and life styles in several of our advanced countries.

For people in many countries, the world of family looks very dissimilar than it was before World War II. Though, the institutions of family and marriage continue to exist as basics of social organizations. In this chapter we shall talk about in detail the concept and meaning of marriage as primary institutions of social life in the Indian context. Although dramatically changes in the character of family can be seen in many societies abroad, mainly Indians hold on to the traditional family life, culture, and traditions. It is bound to stay for a longer period for several more generations to come. So it is apt that we look at the basics of family life and its richness in forming responsible citizens of our society.

THE SOCIAL INSTITUTIONS OF FAMILY AND MARRIAGE

Family is the vital unit of any society. Family consists of father, mother, and children. The stability of family life with parents and children is fundamental to all human society and its extensive-term survival. The sacredness of marriage, its permanence, and its mutuality of love are the major forces for preserving this vital unit in society everywhere crossways the globe. Many discussions are taking place in several societies and cultures in relation to the primacy of individual rights that outweigh any other consideration. The initiatives to promote common good of society based on mutual obligations and less emphasis on individual rights are slowly getting diminished. The procedure is moving towards a total emphasis on individual freedom and choice in this era of consumerism and globalization without due consideration of the negative consequences it can have on the traditional and oldest social institutions of marriage and family. Such a procedure will also have negative consequences on community life and extensive term survival of nations that depend on mutual obligations and responsibilities to ensure stability and peace.

Vital Concepts

Before we proceed with further discussion on these significant and essential social institutions, it is apt that we describe the vital concepts of family, kinship, and marriage.

Family

Several sociologists have regarded the family as the bedrock of society. According to George Peter Murdoc: 'The family is a social group characterized through common residence, economic co-operation and reproduction. It comprises adults of both sexes, at least two of whom uphold a socially approved sexual relationship, and one or more children, own or adopted, of the sexually co-habiting adults." As per this definition, the family members live together, produce offspring, work together, and pool possessions together. The sexual relationships and reproduction of the offsprings through the male and female adults take place as per the norms of their scrupulous society.

According to Anthony Giddens: 'A family is a group of persons directly connected through kin connections, the adult members of which assume responsibility of caring for children'. The Concise Oxford Dictionary of Current English describes family as a 'set of parents and children' or 'the members of a household, especially parents and their children'.

Kinship

The dictionary meaning of kinship is 'blood relationship'. Kinship can be of two kinds:

- Through marriage relationships and
- Through rows of descent that connect blood relatives such as from the side of the father and mother, siblings, offsprings etc.

So kinship comprises either genetic rows or rows initiated through marriage. In short, family is a group of kin.

MARRIAGE IS....:

- What is marriage? It's a union, first of all. It's a uniting, two hearts becoming one. Marriage is companionship and friendship. Its being together, acting together, doing things together, and not wanting it any other way.

- Marriage understands. It's being blind to the fault of the other. It's being considerate in every way-considerate of the other's time, feelings, and desires. Marriage is caring. It's being concerned. It's going out of your way to make sure the other person is well cared for. Marriage is kindness. It's saying kid words and putting those words into action.
- Marriage is support. It's supporting your mate in their endeavors, in their projects, in good times and bad. It's giving them moral support, physical support, prayer support, all approximately support. It's cheering them on and encouraging them when they feel down. Marriage is stooping to lift the other. It's being strong when the other is weak. Marriage is protecting your mate from harm, physical and emotional.
- Marriage is provision. It's doing your part to make sure the other has their needs met. It's pulling your share of the load. Marriage is hard work day after day. It's going out of your way to meet the needs of your mate in whatever way you can, even if you don't feel like it. Marriage is lending a helping hand.
- Marriage is sacrifice. It's giving of yourself for the one you love. It's being ready to provide up your own ideas or desires to make your mate happy. It's giving and giving and giving again. Marriage is yielding, bending, melting together. Marriage is going the extra mile. Marriage is compassion. It prefers the happiness of another to your own.
- Marriage is provided and takes. Marriage is taking turns; it's not one sided. Marriage is submission. It's giving the other person a chance. Marriage is living and loving and helping one another. Marriage is learning from one another. Marriage is humbling.
- Marriage is being opened minded. Marriage is walking a mile in another's shoes. Marriage is listening and understanding. Marriage is being there for the other in bad times as well as the good. Married love stands through thick and thin, no matter how hot the trials or how hard the test. Married love never loses hope. It's always there, always dependable, always ready with outstretched hands and open arms to take the other in- to love, to comfort, to hold, and to cherish. Marriage is learning to let the little things pass.
- Marriage is communicating honestly and openly. It's being willing to humbly share your heart and deepest thoughts. Marriage is talking, praying, discussing, and agreeing together. Marriage does not let things build up flanked by you through ignoring the other, but rather

finds a way; it creates solutions. Marriage is joining hand in hand, heart to heart. Marriage is detection. It's discovering each other, learning in relation to the each other and all the funny things you say and do. Marriage is a good sense of humor. It's relaxing together, enjoying each other.

- Marriage is respect. Marriage is having faith in each other. Marriage is accepting another for what they are. Marriage is coming to the realization that you are not complete without your other half. Marriage can be the mainly satisfying and strengthening experience you will ever have.

Marriage

The Oxford Dictionary describes marriage as 'the legal union of man and a woman in order to live together and often to have children'. It is a legal union which has the sanction of the society, the state, and the religion (for those who believe and/or practice religion). Marriage is flanked by a man and a woman and not flanked by a boy and a girl. Every society (nation) has prescribed norms which describe adulthood.

In India the prescribed age for marriage is 18 years and 21 years respectively for a female and a male. Marriage is a social institution which allows the married man and woman to live together as husband and wife. It is expected that men and women marry not only to live together, but also to have children. According to Anthony Giddens, marriage can be defined as a socially acknowledged and approved sexual union flanked by two adult individuals.

nterrelationship flanked by Family, Kinship, and Marriage

When a man and a woman marry, they become kin to one another. This marriage bond connects together a wider range of kin people. Parents, grand parents, brothers, sisters, uncles, aunties and other blood relatives become relatives of the partner through marriage. Marriage is a stable and healthy union of bonding flanked by a man and a woman who become husband and wife for the rest of their lives. This marital union guarantees the birth of children and a stable and caring environment for them.

Family relationships are found within wider kinship groups. Usually we find two common shapes of families based on the number of members namely,

- Nuclear family and
- Extended family or joint family.

Nuclear family comprises of the father, mother and children (own or adopted). In extended or joint families we can find more members who are blood relatives such as grandparents, parents, brothers and sister of parents, their spouses, children, grand children and so on. There is a steady growth of old age homes in urban centers in India as a result of two reasons:

- Tendency to have small family norm (of having less children) and
- Due to the migration of children (adults) to far absent places within and outside the country in search of job or to live with the spouse.

CONCEPT OF FAMILY IN UN DECLARATIONS

The Universal Declaration of Human Rights through the United Nations in 1948 was a conquest for humanity. This declaration is based on the dignity of the person, and promotes and defends the respect for peoples of several nations and for every one of their members. This Declaration of Human Rights is the authoritative foundation for all subsequent Human Rights Conventions and documents. The right to marry for any man and woman of full age (adult) is defined as a fundamental human right. The family is defined as the natural and fundamental group unit of society and is entitled to protection through society and state. The opening sentence of the preamble to the declaration states that these human rights are 'inherent' and 'inalienable', and stem from our human dignity. It may be noted that:

- These human rights have human dignity as their foundation.
- They cannot be changed through political will or majority decisions.
- They are natural laws, pre-political and apolitical.
- They supersede national laws and no nation can enact laws that change this.
- This UN declaration is the source of all human rights documents, and protects the natural family based on biology.
- There is a right to marry for adult men and women, and to form a family (Art. 16).
- The family is the fundamental unit of society and is entitled to privileges and political protection.
- The formulations on the family and children's rights are repeated in legally binding conventions such as European Court of Human Rights (ECHR1950), and the International Covenant on economic, social, and cultural rights (1966).

Family is the pre-eminent, mainly favorable, and irreplaceable place for the recognition and development of a person on its way to complete dignity. It is in the family that the first steps in human development are initiated and the procedure of education and promotion of human being starts. An individual who does not receive this initial orientation in the family will be greatly hampered in achieving the human fullness to which he/she is described as a person.

Family is the foundation of society. Respect for human rights is necessary for human development of persons in the society. These human development values contain life itself, health, knowledge, work, the community, religion, and culture. Family is the smallest community of persons. The values essential to the family can only be achieved when a man and a woman provide themselves to one another totally in marriage, accept a community of love and life, are willing to fully accept the gift of new life in procreation and accept the responsibility of educating the offspring.

Parents provide the child a home in which the child can grow and develop. All the rights that are necessary through nature for the development of the person in his/her wholeness become real in the family in the mainly effective way. The family, through its very nature, is a subject of rights, the foundational element of human society, and the mainly necessary force in the full development of the human person. The importance of the family's social mediation is undeniable. Family is the 'sanctuary of life'. It is more than any other human reality, a place where an individual can exist for himself/herself through the gift of self.

If the family is protected and privileged through human rights, they are valid in all places and at all times. The survival of society depends on the strong institutions of marriage and family. When the traditional family structure disappears more and more from society, people will no longer know what they are talking about. So the family should not in any way lose its privileged political support. If it loses it, society will be eroded from within. The historical concept of family should not be allowed to disappear. If it does, it would be hard to sustain this existing concept for longer.

There are attempts to redefine the concept of family at national and international stages. There are also initiatives in the legal juridical as well as parliamentary arena. Such efforts also draw power from public debate. It necessity be understood that political majorities cannot change human rights. They are inalienable and inherent. These rights belong to the human being because of his/her dignity and not because some political body granted them.

The UN Convention on the Rights of the Child, 1989

The preamble to the convention states:

- Influenced that the family, as the fundamental group of society and the natural environment for the growth and well-being of all its members and particularly children, should be afforded necessary protection and assistance so that it can fully assume its responsibilities within the community.
- Recognizing that the child, for the full and harmonious development of his or her personality, should grow up in a family, in an atmosphere of happiness, love, and understanding.

As per the UN Convention, children have the right to know and live with their biological parents. If this is not possible, they have a right to adoptive parents who resemble the natural parents to the extent possible i.e. a mother and father. Further, the best interest of the child shall govern all issues relating to children's rights, and not the interest of adults.

Both the UN Convention on the rights of the child and the UN declaration of human rights speak of the importance of the family. The family consists of mother, father, and children. It is a natural institution in all societies. The society is obliged to support it because it uniquely brings forth and rears children. Mother and children have a right to special political support. The pregnant as well as the nursing mother is vulnerable, yet provides the mainly significant work in any society. The family is politically relevant because it is here that the humans are born and raised. Parents do the key work for society through bringing up and educating children to be responsible citizens. No one can replace the parents. That is why we say that the family is the vital cell of the society.

Article 3 of the Convention states that 'the best interests of the child shall be a primary consideration'. This legally binding Convention leaves no doubt in relation to the right of the child to his/her parent, not only to know them but also to be raised through them. This is what is meant through the 'best interest of the child'. It also implies that it is an injustice contravening international law, to conceive a child with the intention of raising it alone. There are instances where hundreds of thousands are born and left in the street to fend for themselves children born out of wedlock, co-environment, and through sex workers. It will not be in the 'best interest' of the child for HIV infected to go for a child when the health of the parents themselves is already in danger. Likewise how can the 'best interest of the child' be protected when some countries allow people living in same sex relationship to adopt children.

Attempts to re-describe family

There are sure pressure groups and individuals in many countries who seek to redefine the concept of a 'family' to contain (with equal legal status as the traditional father-mother-children family) same-sex unions with or without adopted children, single parent with own or adopted children, children born out of wedlock (and not widow/widower), live in couples or co-environment and even to the extend of adoption of animals. Several of them tend to accept variations within family structure adultery, incestuous sex, pre-marital sex and other shapes of partners of sex for the sole purpose of achieving self-satisfaction. No doubt, these are all confusing situations to the man in the street.

Social legitimacy does not necessarily have anything to do with legal legitimacy. Minority groups should not be allowed to undermine core of human rights through themselves. It should be equally obvious that no political majority or minority can change the rights of the mainly vulnerable in the society, i.e. the children. It is not appropriate to agree to the view of law in some smaller countries-that law necessity reflects the reality of social practices and majority will on every issue. For instance, it is estimated that in excess of 50 per cent of Norwegian children are born out of wedlock. Statistically this fact can become the norm. In fact there are arguments through pressure groups in some smaller nations that some of the international rights contradict their national public opinion and practices.

Children and their 'safe haven', the family, necessity not be manipulated in political and legal procedure for their own sake and the sake of safeguarding international human rights. There necessity is no confusion in relation to the family is, and in relation to the children's rights. Children have the rights that adults do not have and these rights precede the rights of adults. A shift in political focus absent from adult 'rights' to marry and form a family, to children's rights to parents and stable family life would also entail that the extreme individualist focus would have to be toned down.

FAMILY LIFE IN INDIAN CONTEXT

Society is an aggregation of small communities of families which live together and co-operate with one another to ensure their own growth and survival. Family is the smallest community of people consisting of an adult male and a female who make free agreement through the institution of marriage to live together, have children, rearing and educating them. These are also functions for which parents need the help of other families particularly from the kinship circles and the society at large.

The traditional Indian family still continues to be a 'joint family'. When sons and daughters (matrilineal) get married, they do not leave their family. They continue to live in it with the spouse and children. The joint family works as a clan in which each member knows his/her role and count on the -solidarity and support of other members in times of need. The father/mother (as the case may be) gets the respect and obedience from children and grandchildren. His/her power reigns supreme. Although majority of the families in India belong to the matrilineal pattern, there are sure groups such as the Nairs in Kerala and the Khasi in Meghalaya who traditionally follow the matrilineal pattern.

In mainly Indian homes, parents are responsible for finding a partner for their child. This is usually seen as a sacred duty. In very few situations, do young people find their own partners. Mainly often such marriages are found more in the middle of the educated urban dwellers than the less educated ruralites. In arranged marriage, the parents take into consideration family background, financial, social, religious and educational status of the would be son/daughter in law. Thus the marriage involves the full participation of two families and children usually accept the choice of their parents with trust.

With the power of better education and job opportunities and exposure to electronic media, youngsters do not accept blindly the power of their parents. They are searching for more freedom and independence. Concerning selection of partners, young people of today tend to demand more freedom to express their desires and to make their own decisions. With the fast changing social scenario as a result of industrialization, urbanization, consumerism, exposure to electronic media and availability of better professional and vocational education, the pattern of family life is also undergoing changes. Such changes are also affecting towns and villages crossways the country. With these changes, a new kind of family life is emerging which is more self-governing and striving to become more self-sufficient.

More women are getting educated. More women are also taking up jobs outside than within family set up. Several of them prefer to marry at an older age than in the past. Women are also becoming more conscious of their dignity and want to be more self-governing and responsible in the family and society. They want to have a say in the matters of the family and decision making and often extensive for a career outside the family. In fact large number of legislations is being enacted to improve the social circumstances of women in India including their political participation at local, state, and national stages.

Value System

India has a rich value system. This traditional value system is still significant and should not be allowed to die out easily. Although there are pressure groups and issues associated with consumerism and expansion of electronic media which power the lifestyles of people, we should not let ourselves be overtaken through such powers.

Much havoc was done in the western countries, Americas and Australia through the liberation ideologies of 1960s and 1970s which advocated freedom of sex, individual satisfaction in excess of common good, feminism and the challenges on the very subsistence of traditional family system. As a result several societies crossways the globe faced the breakdown of traditional family system.

New models began to appear such as single parent families, people living together without marital contracts, cohabitation, same sex unions, divorced and re-married with children of previous marriage, grandparents living with grant children, emergence of old age homes, orphanages and destitute homes, addiction families, expansion of sex industry, sex shops, pornography and sex tourism. All these initiatives have caused much damage to the rich old customs and practices, morality, principles of life, character, and value system.

New models of family life and relationships are necessary to meet the needs and expectations of new generation. It may not be wise to confront the youth of today with ready-made customs, practices, and family patterns of the past. Though, it is very significant and necessary to educate them in relation to the positive characteristics of decent life, morality, brotherhood, regard for parents and elders, correct meaning and purpose sex, marriage and family life. This will enable them to find new ways to practice the traditional values in harmony with the modern times.

Trends Challenging Values

Global society has been facing tremendous threats on its value system from sure groups, events, and developmental initiatives which have far reaching impact. It is a matter of pride that through and large the Indian democracy has supervised to withstand such threats and preserve its rich values and culture and traditional practices. This vast democracy of in excess of one thousand million people is proving to be a model and hope for the rest of the world as far as preservation of value system is concerned. More and more women are getting educated, taking up jobs, and involving in the political and socio-economic development of the country.

Certainly the women empowerment procedure also has brought in relation to the lot of positive and negative trends in the country. Yet the overwhelming majority continues to remain stead fast in their lifestyle and preserve the family values, moral values, and religious practices. The advancements in science, education, media technology, and power of the western life-styles have made some impact on the Indian society. As a result sure unethical practices which challenge the age old value system also have crept into our society. This has increased the percentage of pre-marital and extra marital sex, teenage pregnancies, and abortion, prostitution, mental illness, drug addiction, and divorce, several sexual orientation, single parent households, suicides, and demand for euthanasia. Although all these are part of today's society, the power it makes through the media tends to corrupt young people which are thought to be disastrous for the society. In order to counter this deteriorating trend, there is a need to educate people in relation to the consequences of such social troubles. It is also necessary that we record some of the fading values which will serve as a reminder for the younger generations to come.

Family is Universal

Family has been seen through sociologists as a universal social institution. It is an inevitable part of human society. Usually family has been regarded as a good thing for the individual as well as for the society. For the individual, family life is a necessity for survival, growth and a meaningful life in the society. Survival and sustainability of the society depends on the humans born and brought up as good citizens in the family.

So any consideration for evolving alternatives to family would not be desirable. We have seen the women's liberation movements which began in the 1960s in Europe and America. Some of the feminist writers went to the extent of condemning family as an institution. Although such initiatives shook the foundations of family and its traditional view, hardly any positive outcome was noticed. In fact the contribution of such initiatives brought in relation to the much chaos in the society which witnessed the augmenting divorce rates, decline in moral values, destitution of the aged and children, sexual promiscuity and 'individual thoughts' in excess of 'common good' of the society. To sum up, break down of family leads to the break down of the society and continuation of human survival.

RELATIONSHIP AND BONDING IN FAMILY LIFE

Relationship and bonding are central to family life. Change in family and household composition has brought in relation to the shift in expectations of people from their relationships. Today relationship has become something active. One has to work it out.

It depends on winning the trust of the other person. In fact sustainability of mainly sexual relationships and marriages in urbanized countries depends on this shift in expectations from the partner. Relationship depends on the extend of communication and collaboration flanked by people.

Emotional communication is part and parcel of sexual relationship, love flanked by spouses and flanked by parents and their children. Although much transformations have taken place in urbanized nations, the traditional Indian Society has not changed much.

There is much bonding and very healthy relationship existing within families and flanked by families both in urban and rural areas. Though, one can also observe the power of the western culture or life style gaining momentum in the Indian sub-continent.

The mainly significant recipe for the success and sustainability of marriage and family life is healthy relationship and to be able to bond permanently. The ability to bond depends on the early experience of reliable love.

A child who has never doubted the commitment of both the parents to each other and to himself/herself will naturally develop trust and openness. In several parts of the world there are attempts to present sex as an action with little regard to the relationship. Such attempts offend the dignity of the human being. In fact it strikes at the heart of marital relationship. In course of time it will undermine the family and ultimately the society.

Family life education is planned to help young people in exploring the meaning of human person, expressing oneself through a unique body in its intrinsically masculine and feminine form. Through one's body a person can live out to the fullest his/her highest calling: "To love".

A family cannot come into subsistence without biological union that results in offspring. Only a female can conceive with the cooperation and sharing of a male. A child necessity necessarily has a father and a mother and a home to grow.

This is the fact and the experience of all societies from time immemorial. This concept presents sexual union in marriage as an extraordinary act of total giving of self that has the potential to co create a new human being. A true relationship leads to a meaningful and fruitful life.

Motherhood, Fatherhood and Parenthood

We know that natural law and divine nature assume the subsistence of a fixed human nature, consisting of two sexes (eunuchs are exceptions), where family is a natural and constant institution in human and social life. Motherhood and fatherhood are so constants and the family cannot be re-defined. It exists as a norm in all societies except in cases which differ from this norm due to widowhood or single parent families as a result of divorce or separation. In fact motherhood and fatherhood exist as archetypes of human subsistence with much more than mere biological qualities.

There is no recognition of what a mother is unless one has personal experience of a 'real mother' naturally in relation to her children. Likewise fatherhood can be recognized only through lived experience in a family setting. Parenthood is dissimilar from motherhood and fatherhood. There are several child fathers and child mothers. Human being is the only living creature that lacks built-in mechanism to guide mainly characteristics of one's behaviour. One has to use relationships with others in order to develop talents, skills, and ideas. This is an essential condition for growth towards personal maturity. Family is the first environment which helps a person to relate oneself to others with natural love and cooperation. With the entry of a person of the opposite sex into one's life, a new kind of relationship and experience is urbanized which finds fulfillment in marriage. Together they learn to leave behind the 'I' and 'you' and think of themselves as 'we'. A trust and understanding grow. They become more open to each other, revealing and sharing more of their inner being, feelings, values, and choices.

Every person is meant to be born in a family and described upon to build a new family. Only through accepting family life and its responsibilities seriously a person can learn to reach out to others in love, respect and with a sense of service mentality. The joys and worries of planning, conceiving and bringing up children and social tasks and duties of family, compel the couple to forget self and become more involved with others. So we may say that a responsible parenthood is the real parenthood. One has to be responsible to oneself, family, spouse, children, and the society.

Culture of Life

There is a need for a strategy and rationale for promoting family life because of the new trend which calls for greater 'individual rights'. There is a demand for individual right based on privacy, convenience and some even go to the extend of demanding the same under fundamental rights. The call for 'rights of individual' overshadows concern for common good.

Denying the rights of the terminally ill to choose death in excess of life is not inhuman. The Indian constitution and legislations do not permit euthanasia. Likewise protecting the 'right to life' of an unborn child should not be measured as violence against women. It is the duty and responsibility of parents and young people to protect human life from conception to natural death. In abortion a weak and defenseless child is brutally murdered through those who are supposed to love and care for the child mainly: the parents and health care professionals.

Several young people are unaware that at conception a new life is shaped. Conception takes place at the moment of fertilization: i.e. the moment when the father's sperm and mother's egg cell meet. People with vested interest may make several interpretations. Scientifically one thing is clear: conception refers to fertilization and not implantation. Implantation takes place in relation to the five days after fertilization. The zygote starts to grow as any other living being. It is human because the chromosomes combined at fertilization are that of the human father and human mother and not of any other living being. There are many anti-life forces at work in the world who try to put forward justification for violation of human rights through them and rationale for their thrash about on preference for individual rights in excess of common good.

It is so apt that we teach the young people a culture of life in the face of such public opposition through false propaganda in relation to the individual freedom to abort or to take life (euthanasia). Freedom is a divine gift and to live to its demands fully is our responsibility. There is a tendency of co-environment in the middle of some people in the urbanized countries. Such trends have resulted to some extent in replacing marriages, rise in the rates of divorce, augment in abortion and the number of children born out of wedlock. Some nations have legalized same-sex marriages. In such societies there are pressure groups to recognize homosexual marriages and family status for people living in same-sex unions so that benefits exclusively meant for natural and traditional families are also made accessible to homosexual couples.

Homosexual act under section 377 of the Indian Penal Code (IPC) is a criminal act as on date. A recent Public Interest Litigation filed in Delhi High Court through an NGO to repeal this section was rejected. The Law Commission of India observed that Indian Society in general disapproved homosexuality and this was strong enough to justify treating it as a criminal offence, even if adults indulged in it in private. Throughout the proceedings it was stated: while a government could not police morality in civil society, the law had to express and reflect public morality and concerns in relation to the harm to society at large.

If this was not observed, whatever little respect for law was left would disappear, as the law would have lost its legitimacy. It is a common knowledge that same–sex unions are contrary to nature. It is sterile concerning new human life. It is against morality. A moral environment is necessary for maintaining a healthy society. Though, it is a fact that there are several sexual orientations in the middle of people of several cultures.

Our objective is not condemning such behaviour but understanding how society perceives it. As a result of lifestyles which promote divorce, same-sex unions, euthanasia and escalating rates of abortion, the bedrock of family is being shaken. There are less number of children in families and an augment in aging population which will not be supported through young people.

In order to counter such negative trends and to preserve a value based society, there is a need to promote a 'culture of life' and not a 'culture of death'. We need to have a culture of life in which virtues are cultivated. The ability to bond requires a culture which allows the young man and young woman to wait until a commitment is made before embarking upon sexual intimacy. The wounds and emotional scarring caused through the casual experimentation into which young people are led deprives them, to a great extent, of the possibility of permanent bonding. Commitment and waiting for the right person to whom to commit are at the base of a stable and happy sexuality, marital life, and strong family bonding.

In order to preserve the rich heritage and values of the traditional Indian family, there is a need to collaborate with broader community for stable and healthy family life. There is also a need to encourage governments and several social systems to have an open policy to promote stable family life keeping in view the needs of the society.

FAMILY LIFE EDUCATION— CONCEPT AND MEANING

India has a extensive tradition of closely bonded family system. Today, though, some people argue that [modern] family as an institution is facing crisis. Let us take a more optimistic view. Families cannot be in accessible from the impact of rapid social and technological changes, yet, they have ably demonstrated to turn difficulties into challenges through offering support and security to each of its members in a fast changing society.

Each family is recognized with a unique sense of shared family traditions, shared experiences, and stability of behavioral patterns through generations. We believe that nurturing this excellence of family get in touch with and interaction becomes even more significant in a time like today when several people feel lonely and, rootless. Technological progress often contributes to development but sometimes it generates a sense of depersonalization and deprivation from one's own identity. That is why family life education is essential and relevant in the present day context.

Every society in excess of time has evolved its own ways of preparing its younger members for adulthood, primary through an educational procedure. Traditionally, mainly elements of family life education have been informal, taking place within the home, at place of worship, at work and in day to day contacts with other people. In childhood, several values related to family life education have been imparted through stories from folk tales, epics, religious scriptures, etc. The hidden rationale behind these is to enable children to imbibe the right values for family life or to inculcate these values in them so that these act as guidelines for the rest of their lives.

In tribal communities, initiation ceremonies are performed which symbolize the entry of a child into adulthood. Thereafter, the child can have access to the knowledge and privileges that are permitted only to adults. On a similar note, in Southern India and several other parts of India, a ceremony mannered at the age puberty introduces a girl to all the intricacies related to family life education. So, through the procedure of socialization, through observing people's behaviour and through their own experience, children and adolescents become familiar with the norms, customs, and values of their own society. These experiences give the guidelines for their own behaviour.

In the present context, families in several parts of the world are finding the task of helping young people to prepare for adult life increasingly hard. Often it is found that they have as little awareness as their children in relation to the troubles they are facing. It is here that the role of schools and of voluntary organizations becomes significant. These institutions replace traditional form of education and offer guidance to people to enable them to adjust to changes. The development of organized programs of Family Life Education is one way of showing their willingness to work together with families, to assist young people in their transition from childhood to adulthood.

Very often it is observed that family life education is treated synonymous with sex education. In reality, though, family life education is a much broader concept. Sex education is only one of the components of family life education which also covers economic characteristics, social characteristics like marriage, responsible parenthood, besides characteristics of sexual health. We will talk about all these things in detail under section 1.3. Like sex education, family life study is dissimilar from family life education.

DEFINING FAMILY LIFE EDUCATION

Several international organizations and eminent persons working in the area of family life education have tried to describe it in several ways. Some of the definitions are given below:

- "Family life education refers to those educational concepts and experiences that power attitudes towards family living, personal relationships and sexual development"
- "One comprehensive and attractive approach perceives family life education as catering for individual needs leading to personal growth and enabling the individual to function as a responsible member of the family and society".
- Family life education comprises a study of self awareness, understanding of others, of sexuality, marriage, and parenthood. The

knowledge gained and skills urbanized will contribute to the individual's ability to cope both with social change and with relationships in society as a citizen, spouse, and parent.

- Family life education may be defined as education for human development which seeks to ensure that each individual approaching adulthood is equipped with the skills and personal reserves to cope with the challenges of every day life in society within acceptable societal structure and to adapt to change with experience and equilibrium But none of these definitions seem to cover the whole aspect of family life education. The definition given through International Planned Parenthood Federation is closer to what we usually consider as family life education.

All these definitions, though, ignored the economic factors involved in family life education. So, in this chapter, we would like to describe family life education in the following way:

- The concept of family life education refers to a variety of formal and informal efforts through which persons become ready for the roles and responsibilities of family life.
- Rapid technological and social changes of today's world have increased the need for individuals, families, and societies to enhance interpersonal and decision-making skills of each member of the family, especially those of the husband and wife, and reinterpret the meaning of mutual commitment, as well as uphold support for their economic self-sufficiency and emotional stability.
- Rising rates, marital breakdowns, family violence, falling indices of marital satisfaction and the time family members spend together suggest that people are either inadequately prepared or lack support in facing the challenges of marriage and of family life.

Yet the sheer number of stable, satisfying, and rising marriages, especially, within the nuclear families, in the midst of personal and social pressures, and inspires of limited preparation and support, suggests that several are seeking to build strong families. Likewise, improved relationship skills in the middle of the educated at high school, college, and in community setting, perhaps signal a rising desire to invest for the success in marital life through trying to avoid possible distrusts that are likely to arise within relationships.

Meaning of Family Life Education

Family Life Education is a broad and flexible field. Anything which contributes to the total growth and well being of the family – physical, mental, emotional, economic, and spiritual – can be incorporated under the umbrella of family life education. That is why family life education has its roots in several disciplines including sociology, social work, psychology, anthropology, biology, education, history. The goals of the programme are often broad based. In excess of all its objectives are to promote the freedom to choose parenthood and enrichment of human life.

Family Life Education is measured as a value related concept. Mainly of the values related to family life education are deeply rooted into the socio-cultural milieu of the people. The values are moral or ethical, cultural, religious, personal and social.

CONTENTS OF FAMILY LIFE EDUCATION

The content of Family Life Education is related to the needs of those for whom these are meant. It is vital to ask young people while preparing them for adult life concerning the kind of educational input to be provided to them which will help them in taking decisions and developing their skills. This brings us to analyze several subcomponents of Family Life Education:

Family: Meaning, Kinds, and Functions

Family is the vital unit of society. Although social scientists have studied a great deal in relation to the various facets of family structures and organization, they do encounter difficulties in developing a broad, culture free generalization that would be applicable to families in a wide range of societies. One major difficulty arises from the troubles of definitions associated with the concept of "Family" itself. The significant question of what constitutes a "Family" is not easy to answer because through itself the term "family" is ambiguous. To describe more precisely the concept of family, the kinds of families have to be examined.

Basically there are two kinds of families: the nuclear family and the extended family (joint family). In extended family, parents, grand parents and even great grand parents live in the same home or in neighborhood. The family members of Nuclear family, consists of a single family of husband, wife and children. This kind of family is mostly found in urban and industrial societies. In some countries, young couples in a nuclear family setting are at a loss to whom to turn for advice in the absence of parents and elder relatives. Behaviour scientists refer to these as family discontinuities (e.g. first conception, birth of first child and so on.), family crises, and other difficulties. In terms of functions, the strength and solidarity of society are highly dependent on how the family as a vital unit performs its vital functions. It is the responsibility of all members of a family to fulfill family functions, although parents usually shoulder a superior share of these functions. Failure on the part of the parents to perform their duties can lead to social troubles for several generations to come. Social troubles in the middle of children often can be traced back to the failure of parents/ families to perform their functions. To have a truly happy family, the needs of each and every member necessity be met. These contain the provision for vital needs of a person, which according to the Psychologist Abraham Maslow are: physiological, safety, love and belonging, self-esteem and self actualization.

Family Roles, Relationships, and Responsibilities

People of other countries often express a surprise at the roles and relationship followed in an Indian family. They are surprised that in Indian Family set-up, children below ten years of age too are expected to play significant roles in socio-economic activities such as taking care of cows and buffaloes, collecting cow manure, fetching drinking water, washing clothes, taking care of their siblings and so on. It is also inconceivable to several of them that married children sometimes remain dependent on their parents and that aged parents too are dependant on their children.

In Indian settings, traditional family relationships are usually quite extensive. In addition, there are other kinds of family relationship, such as god-father god-son relationships and relationships arising from a network of inter-marriages flanked by families. For instance, family ties are quite complicated as these are based not only on blood kinship but also on past associations as school mates, co-workers and so on. In short, family relationships are governed through several customs and traditions which are more binding than rules or laws. A clear understanding of roles in the family is very significant. Misinterpretation or lack of understanding lead to troubles and sometimes to chaos. For any family to exist, there are sure tasks which necessity be performed through its members. The issue of roles within a family is further complicated through changes in the family system, that is a change from extended to nuclear/ conjugal families. The participation of women in economic activities out side the home, the power of western culture, access to higher education, opening up of the economy in general, and related activities are affecting these changes.

Family Life Cycle

A study of family life cycle gives a foundation for the study of the composition, growth, and development of families. An understanding of the stress and strains experienced at each stage of the cycle paves the way for better understanding and co-operation in the middle of family members. There are basically eight stages in family life cycle.

- Stage I: Bearing families: At this stage, a couple learns to synchronize its ideas, habits values and so on. The married partners are learning to live with each other for the first time. Differences are found to appear from time to time and it is significant that a couple irons them out and establishes a meaningful marriage.
- Stage II: Child bearing families: The couple enters this stage at the time of first conception. With the coming of a child, finance, leisure time, entertainment, and privacy will undergo changes. Proper child

care becomes very significant at this stage. Attention is diverted or shared flanked by the newcomer and the spouse.

- Stage III: Families with pre-school children: At this stage, parents need to cater for the critical needs and interests of pre-school children to stimulate their growth and development. Through this time parents think in relation to the question of having or not having more children. The energy depletion in homes where domestic help or help from relatives as in extended families is hard to come through is one major are a of adjustment. It is common for women to assess that the demands of home keeping, wage earning, mothering, and being a wife are overwhelming on them. Husbands cannot afford any more to assume the traditional role of wage earner only. They need to share the household work, and some may find this to be hard. Misunderstanding flanked by the couple can also result when too much attention is given to the children.
- Stage IV: Families with school children: Families with school going children have to pay attention to the child's education and growth needs. The parents are also expected to collaborate with the demands and expectations of the school in helping the child in his/ her studies and in development characteristics.
- Stage V: Families with adolescents: Adolescents have special troubles that need the help of parents and this stage can be very crucial for them and their children. As the teenagers approach adulthood parent's necessity ensure a balance flanked by freedom and responsibility through helping their children to plan and then to execute the plans which will be beneficial and productive in several life situations.
- Stage VI: Families as launching centers: After the young adults have gone into work or studies, parents begin to feel a sure sense of loneliness. It is often referred to as 'the empty nest syndrome'. At the same time they also have the need to uphold a supportive home base for the children. Children need support and guidance for several cultural, religious and traditional family based events like marriage, birthday celebration and other ceremonies in the company of parents and relatives.
- Stage VII: Families in the middle years: This is a trying time for a couple who have to adjust to a dissimilar life again absent from the children and absent from work. It is significant at this stage that hobbies are urbanized/or urbanized earlier in life can stay their occupied throughout the time after retirement. The reduction in the income makes it significant to adopt a more modest standard of living.
- Stage VIII: Families in the late years: The family members need to learn to cope with bereavement and living alone at this stage. Society's expectations will stay children absent from the concerns of aged

parents. Very often helpers like home nurses may have to give the required services to the parents.

Family Possessions

Family needs are related to the vital human needs as described through Abraham Maslow in his Hierarchy of Needs. According to this theory all human beings have five vital needs. They are:

- Physiological needs – food, drink, sex, clear air, and good health
- Safety needs – to have a roof in excess of one's head, housing, clothing, etc.
- Love and belongingness need – to have a family or community to belong to, to have shoulder to lean on, the need for acceptance, giving and getting love.
- Esteem – Self respect and respect for others.
- Self actualization – Self fulfillment and reaching one's potential, to become somebody in life.

This need hierarchy theory can be graphically represented in the form of a pyramid. Needs at the lower stage are to be met before the higher stage needs.

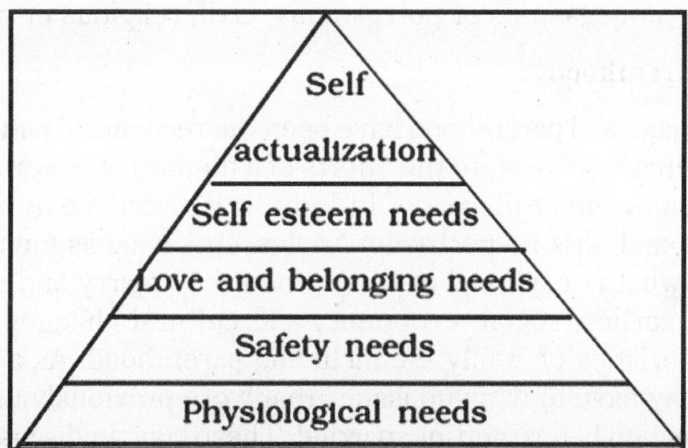

Family needs are unlimited and possessions are limited. Family or individual will use human and non-human possessions to meet the unlimited needs. A family has to manage the use of its possessions in order to maximize satisfaction derived from them. Family possessions are discussed here in the context of several stages of family life cycle and family size. It should be borne in mind that every event in dissimilar stages of family life, drains family possessions. The demand on the family possessions is heavy, especially when there is an overlap of dissimilar stages. It is significant, for any newly wed couple to decide when to have the first baby, how several children to have in all and how they should be spaced. Family size affects the need satisfaction of the family members, as mentioned above. Every child has the right to have a balanced diet, adequate clothing, safe shelter, proper education, attention and affection, and medical attention as well as the right to meet all the vital physical, mental psychological and spiritual requirements of a healthy and happy life.

Marriage

Courtship and marriage are two issues which mainly adolescents begin to be pre occupied with. Some adolescents are forced into marriage because of pregnancies; others marry to escape family pressures, and a few others want to escape from schooling while some others get married because of the prevailing custom. Appropriate educational programmes are needed to prepare young people for marriage. After all, marriage in India is measured as a permanent relationship.

While discussing marriage, the several questions concerning reasons for marrying are raised. For instance, what are the mainly common kinds of marriages? How do people meet their marriage partner? How extensive does a couple usually knows each other before marriage? What is the mainly common age for marriage for a man and for a women? Is dowry or bride payment a good thing? What are the usual marriage ceremonies? In India, marriages are usually classified as follows: Marriage through free choice of the partners or love marriage, arranged marriage and forced/arranged marriage. They can further be classified as monogamous or polygamous; civil, religious or customary.

Responsible Parenthood

Early marriage and parenthood have been the recognized pattern in several societies and remain so, despite the efforts of a number of governments to rise the legal minimum age at marriage. In India, minimum age at marriage is 21 and 18 for boys and girls respectively. Adolescents need as much information as possible on what is expected of them when they marry and raise a family. As mentioned earlier, social, economic, and cultural changes are affecting several characteristics of family life including parenthood. As a result, young people today may have to wrap up issues that were previously unnecessary to take into explanation before getting married. These contain discussions through both partners on the desirability of having children. If children are desired, will the father help in caring for them? Will the mother be allowed to work outside the home? Who will manage family funds? Becoming parents brings responsibilities which expand in excess of the years. Having children is measured to be a natural pattern of family life and some couples do not seriously weigh the implications of parenthood.

- To avoid the risks of hunger and financial insecurity, parents should plan the number of children, based on their ability to support and rear them to full maturity.
- To reassure themselves of the benefits of parenthood in their old age, parents should give their children with guidance and direction

so that they may develop and inculcate the right values as they approach adulthood.

- To help in achieving an orderly society, parents should strive to bring up a family whose members are cognizant of both their rights and duties, while recognizing the benefits of the society as well as the tasks of supporting it.
- Parents who respond properly to their partners' personal needs are, in effect, reducing (if not, eliminating) the probability of their ever becoming estranged.

Responsible parenthood cannot be discussed without bringing up the issue of family planning. Family planning is a means of enhancing the excellence of life of families including regulating and spacing child birth, helping fertile couples to beget children, and providing counseling for both parents and would be parents.

TRADITIONAL INDIAN VALUES RELATED TO FAMILY LIFE EDUCATION

In India like some other Asian and Pacific countries, where status of women is lower than that of men, socio-cultural values like preference for son, early marriage, are quite apparent. Continuation of family lineage, social security that comes from a cohesive family cohesion, performance of religious rights for salvation of parents' soul after death in accordance with sure religious traditions (especially the Hindu religion), making free labour and economic benefits accessible to parents in the form of increased income and dowry, are all significant motivations related to family in our country.

In rural areas, children help parents in planting and harvesting of agricultural products, cooking, cleaning, fetching water, and looking after cattle and younger ones in the family. The phenomenon of child labour (though illegal) prevalent in unorganized sector, supplements the income of poverty stricken home- holds. Still in rural areas the girls are encouraged to marry young and to have children at a relatively tender age due to prevailing social customs, in spite of legal provisions concerning the minimum age of marriage. Though child rearing at an early age is socially acceptable, the practice several restrict women from educational, social and economic opportunities, Early marriage and unplanned birth of children are found to be responsible for ill health of both mother and the child. In this country child mortality due to several diseases is very high. The traditional joint families with their numerous advantages are on the decline. In both urban and rural societies the birth of a child is measured a blessing of God. Some of these values are deeply rooted in our culture and it is hard to change them in the close to future.

Marriage is deeply rooted in socio- cultural ethics. The traditional value of an Indian family is that a girl and a boy having somewhat similar social status bond letter in a marriage. Stage of income, of education and the question of compatibility flanked by the two are of a secondary importance. Usually the boys are a couple of years older than the girls especially in the case of arranged marriages.

In urban areas, the prime consideration in the marriage of a girl now-a-days is the income of the boy and economic status of the parents of the boy. Traditionally it was taken for granted that marriages should be arranged with the consent and advice of the parents.

In several middle and upper middle class orthodox families horoscope matching before marriage is mannered before performance of any religious ceremony. In traditional Indian society, dowry system was not practiced, but the system of dowry is increasingly practiced in the context of changing value system of middle class and of high society.

Though dowry has been legally abolished, it still does exist with added vigor, making the life of the parents of girls and the life of the girls themselves miserable. At times attempts are made to justify dowry as the right of the girl to the parents' inheritance. In fact, dowry is often given from what the parents do not have: borrowed money.

In traditional Indian society, inter-caste marriage was discouraged. Child rearing was measured as the principal responsibility of the wife. Decisions in the family were the prerogative of the father of the household. Traditional Indian society could be measured as a very reserved society so far as dissemination of information concerning sex role/sex relationships and sex responsibilities are concerned. Sex was approximately a taboo subject. Male children have always been given preferential treatment.

The wife was blamed if she could not deliver a male child. Home was measured as an ideal place for women and women were discouraged from working or looking job outside the home. But in urban society, the scene is changing fast. In India it is believed that if women were to take up a job outside the family, the family would be neglected.

Women from poor families are occupied to do all sorts of work for others. Education for girl children is discouraged. Division of Indian society into several castes and religious rituals was strictly followed.

Strict discipline was enforced through the parents for the development of the children. Although, the power of some of these values has changed in urban areas, they still play a major role in decision making in mainly families in rural areas.

Reorientation of values on several issues related to dynamics of family life is a extensive and hard procedure but the experience in a number of countries shows that it could be achieved.

In all societies, traditional values are increasingly being challenged resulting in disagreement flanked by adolescent children and parents. In ancient India, male members of higher caste were advised to practice Purusarthas and

Ashramas. Through which each individual passed were directly related to family life.

Purusarthas

Purusarthas represent the fundamental aspiration of ancient sages and social thinkers. These are values or ideals for which one necessity strives for in conduct and behaviour. Hindu social organization is both material and spiritual and these help people to co-ordinate the two. There are four purusarthas: Dharma, Artha, Kama, and Moksha.

Dharma

The word dharma comes from Sanskrit which literally means to preserve' It represents right action. It stands as principle for maintaining the stability of society. It is the guiding principle for the attainment of other purusarthas.

Artha

Artha refers to acquisition of material property. It is significant to acquire material wealth because it satisfies the material needs for running a household, to provide gifts and feed the poor. This should be earned through dharma or right action.

Kama

Kama refers to all the desires in man for enjoying and satisfying the senses including the sex desire through right action. It is necessary to satisfy physical urges of man as well as achieve propagation of species.

Moksha

It is the supreme and final purpose of life and is attained through proper functioning of other purusarthas. Moksha is mainly concerned with the individual and frees him from the bondage of cycle of birth and death.

Hindu Ashramas

These four purusarthas are the foundation of Hindu Ashramas. They are concerned with the link flanked by the individual and the community. They form the psycho-moral foundation of Indian social system. The Ashramas are the four dissimilar stages of life in which specific functions have to be performed. The four Ashramas are Brahmacharya, Grahyastha, Vanasprastha and Sanyasa.

Brahmacharya

This is normally accepted as the first twenty five years of life. In this stage each young man maintains celibacy and stays at the home of the Guru for studies and learning.

Grahyastha

Roughly after that twenty five years of life form the period of grahyastha in which an individual gets married and leads his family life.

Vanaprastha

It is the third quarter of life in which an individual starts dissociating himself from the family life and starts the life of a pilgrim or of a mendicant.

Sanyasa

This is the last quarter of life in which an individual dissociates himself totally from the family and becomes a monk.

IMPORTANCE OF DISSIMILAR KINDS OF VALUES IN FAMILY LIFE EDUCATION

Through now you might have realized the importance role that the value system plays in family life education. Values have been recognized as a theme in family life education and as a problem confronting family life educators. Questions have been raised in relation to the role of values in family life education and in relation to the appropriate responses of family life educators to several values, issues and concerns. What values should be incorporated or excluded from family life education programs? Should educators share their personal values with participants? What is the best way to handle controversial values questions? How should family life educators deal with potential differing values.

Kinds of Values

There are dissimilar kinds of values. Although it is not possible to give an exhaustive list here, let us try to look at some of the mainly significant kinds of values which contain: moral or ethical, religious, aesthetic, health, economic, legal, cultural, educational, personal, and prudential value. Mainly of these kinds of values are relevant to the purposes and subject matter of family life education. One of the major operating principles in family life education is to "respect differing individual and family values".

Special attention needs to be given to moral or ethical values and to the relationship of these values to cultural, religious and personal values. So here we shall talk about only these four values i.e. moral or ethical values, cultural values, religious values and personal values. We will also study how they act as operating principles in family life education.

Moral or Ethical Values

Several family life educators are often uncomfortable with the word moral because the word 'moral' seems to be a relative term. It varies from society to society, culture to culture place to place and from time to time. The words ethical and moral mean the same thing and can be used inter changeable. In literal meaning, these terms pertain to human conduct and character, and usually refer to moral rectitude of an action, whether they are good or bad. Moral and ethical vales are expressed in terms of principles or rules of right conduct. Values such as personal integrity, tolerance for diversity and differences, social responsibility, respect for persons, sense of justice have an significant place in family life education.

The key point of ethics and morality is to go beyond the personal self interest and to consider, equally and impartially, the rights and interest of all involved in a situation.

The intention here is to be able to overcome egoism and be able to see things from other people's point of view too. A second intention equally significant, is to consider whether or not potential harm to some people can ever be justified. Within family life education, there are several opportunities to talk about issues like violence within marriage, intolerance, caste complexes, etc.

Cultural Values

In one sense, every value is a cultural value, because the central way in which one acquires a value is through acquiring a language. Language is a public cultural artifact. In the middle of other things, language is used to express commitment to sure values and it would be hard to imagine a culture that does not contain, for instance, legal, economic, aesthetic, moral and intellectual values. Two significant points in relation to the cultural values are important to family life education.

First, all cultures have some commitment both to a set of dominant ideals and to a set of greatest fears. In general, the relevance and purpose of moral principles in a scrupulous culture is to facilitate the attainment of the dominant ideals (e.g. to attain equality or familial stability) and the avoidance of the greatest fears (e.g. to avoid use or loss of autonomy).

Family life educators who work with multicultural groups will need to be aware of and sensitive to not only expressions of cultural differences but also how these differences represent cultural attempts to accomplish their ideals and to avoid their fears.

The second point is that cultures do not remain stagnant. There are likely to be conflicts within the culture in relation to its dominant ideals and fears. Some of these conflicts will have to do with topics central to family life education (e.g. which ideal is more significant loyalty to family or personal autonomy). Because these intra-cultural conflicts are sensitive issues and often problematic for families, family life education programs do not prepare individuals to deal rationally with these changing ideals.

Religious Values

Although there are several differences in the middle of religions, a number of characteristics appear to be common to several of them:

- A belief in a supreme being or a set of beings
- A set of concepts (usually a very intricate set) that refer to the postulated relationships flanked by the supreme being(s) and human beings;
- A belief in some sort of subsistence after death.

These characteristics have significant implications for family life education.

First it is clear that several moral concepts and religious concepts such as respect for persons will overlap, leading at least some people to believe that moral/ethical values and religious values mean the same thing.

Such a belief results in potential view of what counts as justification of moral principles. Thus some religious hold that moral principles have weight because they are " commandments" of one or more supreme beings rather than because they are part of an autonomous way of viewing human relations and human actions.

Obviously, family life educators will need to be sensitive to these potentially dissimilar beliefs, but it need not result in an impasse where the educator can do nothing. What is crucial is a willingness on part of the educator to be opening minded, and to be both willing and able to set an instance of careful and balanced consideration of issues that are central to Family Life Education.

Personal Values

Much of the attention to values in family life education appears to be directed towards that is usually referred to as developing an understanding of one's own personal values and of learning to respect the personal values of others. Hamm has made an significant distinction flanked by public or social morality and private or personal morality that has relevance for family life education. According to Hamm, social moral judgment has to do with inter-personal behaviours and addresses itself to vital human needs and fears, wants and desires, which are either to be satisfied or avoided and is a precondition for human beings to have a choice at pursuing excellence of life in its several shapes. Some of the vital principles that are necessary for social morality contain:

- Justice as fairness (impartiality) non-discrimination);
- Non-maleficent (restraint from harming or injuring others)
- Minimal beneficence (the moral risk in satisfying their vital needs);
- Freedom (no right to interfere with others) and
- Honest (truthfulness and non-deception).

In Hamm's view, social morality refers to those things that are "good for all" and has as its summary notion 'respect for others'.

Hamm has suggested that although personal morality shares some of the same characteristics as social morality, personal moral judgments address those things that have to do with "my good" rather than with "good for all". The summary notion in personal morality is self-respect. Accordingly, self-respect necessity provides way to social morality. This has significant implications for family life educators who are concerned in relation to the role of their personal values in family life education. Hamm has claimed that this distinction is an significant one because the subject matter and the strategies required for teaching social morality and personal morality are dissimilar. In general teaching social morality requires serious and systematic attention to the rules and the principles of social morality while education for personal morality requires some value clarification.

ADVANTAGES OF FAMILY LIFE EDUCATION

Let us now talk about some of the needs which provide birth to family life education.

- Family life education provides an educational role: The family has always lived as vital social grouping in the middle of human beings irrespective of the variation in its structure and functioning. There is of course no universal pattern, Families may be nuclear or extended, monogamous or polygamous. But in mainly part of the world the structure or functions of the family are in transition and its educational role is one which is dynamic in nature. So everybody should acquire this knowledge.
- Family life education acts as a crisis manager. Adolescents are rising up in a world in which they will have to make more decisions for themselves compared to previous generations. They tend to experiment more with their life, make choices and take risks and learn through their own experiences rather than through those of others. This can lead to confusion, frustration, despair and risk taking of a kind which is ultimately self- destructive. So family life education has an significant role to play in order to help people avoid all sorts of crisis.
- Family life education provides skills for preventive action and knowledge for decision making particularly in the middle of adolescents; Each new generation of children face health challenges, but those faced through today's school age group seem particularly daunting. Children are confronted at an early age through situations that require knowledge for decision making and skills for preventive action. Very often adolescents find themselves under strong peer pressure to engage in high risk behaviour like drug and alcohol abuse and sometimes sexual abuse which can have serious implications on

their lives. These issues should be addressed to the young and adolescents through family life education.

- Family life education helps in understanding one's own role with the changing family structure and functions. The rate of social change in mainly societies needs to redefine the role of adult family members. Where tradition once recognized the norms for family behaviour, parents are now being challenged to re-look at their roles and to accommodate to the demands of changing family structure and functions.
- Family life education helps in proper understanding of family life cycle: The impetus for family life education programme is based on the pervasive nature of the family life cycle. Each new stage of the cycle gives rise to new learning needs. In the past, these learning needs were met through informal learning activities such as reading, seeking advice from peers and professionals (doctors, religious leaders and priests) and reflecting on personal experience. Some of these activities are now being offered as family life education through formal learning programmes.

National programmes of Planned Parenthood for instance have been recognized in a number of countries to curtail population growth and to promote improved standard of living. The belief too, that parenting, skills, knowledge and attitudes can be learned through programmes of education as opposed to learning through observation has given rise to a host of such programmes approximately the world.

Need of Family Life Education

It is clear that family life education is a new field and one that is moving towards the "professional" end of the continuum. Some of the advantages of family life education are mentioned below:

- Family life education is necessary for each individual: Every individual needs to know in relation to the family life education. Every individual spends eight to ten hours in work for earning for living. For this we study a number of courses in excess of several years. But there is no course for the fourteen to sixteen hours which we spend every day with our family. This needs a formal and compulsory orientation for each and every individual. So every individual should be provided a family life education as a life enrichment programme. There is a rapid change in the family structure all in excess of the world from the extended family to nuclear family. But there is nobody for guidance. So there is an urgent need for this kind of education to cope with the emerging trends.
- Professionals working in this area and allied fields need training:

People who are working with NGOs in this area, paramedical professionals and teachers need an orientation on training in family life education.

- The action becomes a full-time paid occupation: There are thousands of full time paid people practicing as Family Life Educators with specific qualification in several settings throughout the world and in India.
- Training schools and curricula are recognized: Several departments and schools have been recognized at undergraduate and graduate stages of family life education and family studies. The first Ph.D programme in family life education and family studies was recognized in 1962 at Columbia University in New York. There is a unit for Family Studies at the Tata Institute of Social Sciences, Bombay.
- Those who are trained, have recognized professional associations: The recognized professional associations are uniformly supportive of the development of the knowledge base and skills of the practitioners in family Life Education and that has played a key role in its definitions.

IMPORTANCE OF FAMILY LIFE EDUCATION

In the last chapter we discussed the concept, meaning and need of family life education. In the concluding part of the lesson we also discussed the advantages of family life education. But the question which arises is: What is family life education and for whom is it planned? Initially family life education was started informally as people's response to a felt need though, informally this education lived in society as part of social life. Ironically, family life education is a field which came into being as a result of public demand although it still has to fight for public acceptance as a field of education.

Some people still hold the thought that families are private and should not be intruded upon while others uphold an self-governing stance arguing that they do not need to learn how to be better parents, partners or family members. They claim to know it all instinctively. But there is a positive change in attitude practically everywhere in the world, especially since the late nineteen seventies. People are finding it a necessity due to rapid social change that has occurred within several sections of the society. So there is a need for equal participation from family and community to make it more effective. Before we talk about the broad objectives of family life education, we should know the operating principles of family life education.

THE OPERATIONAL PRINCIPLES OF FAMILY LIFE EDUCATION

Family life education has sure operational principles. Let us list them one through one.

- Family Life Education is related to individuals in the families throughout the life span, both as individuals and families.
- Family Life Education should be based on the need of individuals in families.
- Family Life Education is a multidisciplinary area of a study and practice.
- Family life Education takes an educational rather than a therapeutic approach.
- Family Life Education should present and respect differing family values.

OBJECTIVES OF FAMILY LIFE EDUCATION

Family life education is concerned with assisting prospective candidates for learning the procedure of living in the family, continuing family life to the after that generation and sustaining the social, cultural, religious and traditional practices and values. The major objectives of family life education are:

- To develop knowledge concerning the concept of family, its roles and functions: Every individual (both young and old) should have proper knowledge and understanding of the concept of family, its roles and functions. This knowledge will help an individual to recognize the several tasks that need to be undertaken for the well being and maintenance of the family.
- Family life education aims at developing the ability to deal with family relationship and responsibility of family life education helps in establishing and maintaining satisfying relationships with members of the family, with friends and with others whom they come into get in touch with in the context of family life.
- Family life education aims at developing life skills to cope with dissimilar situations: Family life education develops the knowledge, values and skills necessary for adult life, marriage, parenthood and participation in the social life of a community. It also helps in communicating effectively with others and in making wise decisions in relation to all matters related to family life, personal relationships and dealing with several members of the community.
- Family life education aims at helping in understanding and coping with changes in one's own life and changes in the society: Family life education helps an individual to understand and cope with changes in the adolescent life and cope with the society in which one lives. Changes in society are occurring all the time and particularly the breaking down of traditional social structures and the changing roles of men and women due to marriage, education, employment undertaken through women and such like phenomena.
- Family life education aims at developing knowledge in relation to the physiological processes in one's own body especially the procedure

of human conception and birth. It also creates an awareness concerning the consequences of adolescent pregnancies and parenting and of alternatives to pregnancies and contraception.

- Family life education seeks to create responsible and useful young citizens. The ultimate aim of family life education is to create responsible and productive young citizens through helping them to achieve the above mentioned objectives.

Specific Objectives of family life education

The specific objectives of each of the components of family life education are illustrated below:

Components Objectives:

- Concept of family a) To identify the historical kinds and purpose functions of a family and its varying structure; b) To describe the term family; c) To identify several kinds of families; d) To enumerate the functions of the family
- Family relationship and a) To explore the concept responsibility relationship; b) To identify rights and responsibilities in the family; c) To understand the factor that affect family relationship; d) To cite common conflicts that arise in the family; e) To identify the causes arising from differences in the middle of individuals; f) To recognize the several tasks that need to be undertaken for the wellbeing and maintenance of the family.
- Family life cycle to understand the several stages of family life cycle.
- Family needs and possessions a) To identify events in the life of a large family and to determine the stages that place heavy demands on the family possessions; b) To understand the vital family needs and to identify the possessions accessible to achieve them; c) To understand the impact of family size on the vital needs of family members; d) To develop skills for budgeting and managing family economy
- Marriage a) To understand the marriage customs, laws, norms and practices under several cultures; b) To understand the concept of marriage and the responsibilities associated with marriage.
- Responsible parenthood a) To understand the roles and responsibilities; b) To appreciate the responsibilities involved in having children. c) To be able to explore and acknowledge the youths attitudes and values concerning parenting; d) To talk about the concept and importance of family planning and the advantages of practicing family planning. e) To identify the dissimilar methods of birth control, their effectiveness, side effects and safety.

IMPORTANCE OF FAMILY LIFE EDUCATION

The following reasons can be offered to indicate the importance of Family Life Education in today's world:

- Expectations of marriage and family life are changing: When marriage was for economic security and husbands rule the roost, the rules were more simple and straightforward. Expectations for affection were minimal as were options outside marriage. Today, love, companionship and equality are very significant, communication and problem solving skills have become even more valuable for happiness.
- Terms of marriage and family are changing: At the turn of the century, the life span was shorter, more of marriage revolved approximately parenting, and couples often drew on the network of family and friend for support. Today, with nuclear families being on the augment in our country, couples are expected to meet their own needs, with minimal extended family support, and coordinate adult growth and roles through 50-60 years.
- Life doesn't prepare mainly persons for new family roles: Couples face more demands and fewer supports, but few persons learn to be mature and be equal partners in families. Disagreement resolution and problem-solving skills are part of several job trainings and volunteer groups, yet couple communication remnants sorely neglected. Traditional education prepares a person for 8 hours of his/her working life. A person is actually functioning as a member of a family with an assigned role for a much longer part of the day. What preparation does one receive for this formidable and significant role?
- Benefits of happy family life and adjustment: Research finds happy families have lower rates of illness, depression, and disagreement in and outside their marriage. Well-adjusted partners make easier transitions to parenting and tend to be more supportive parents. For mainly persons, marital satisfaction is the best index of family satisfaction; Family satisfaction is the best prediction of wellbeing.
- Consequences of marital breakdowns: Whether divorced or living under the same roof and being totally separated from each other emotionally, partners, their children, and their relatives experience emotional and financial stress which may have extensive-term effects. In scrupulous, poverty, sustained partner disagreement, and separation from old friends and family can limit the horizon of couples in stress and of their children. Mainly persons would choose to avoid families where there is no harmony or peace, or take advantage of this situation.

- Benefits of family life education: Researchers have successfully shown the benefits of training programs in relation to increased couples' skills and satisfaction for extensive-term, lowered marital disagreement rates, and reduced family violence rates.
- Starting early prevents breakdowns: Couples who are aware of the principles of family dynamics, of problem solving methods, and of the support and social possessions, become more aware and more capable of handling the challenges of family life. Proactive couples tend to avoid relationship-treating crises and enjoy the benefits of life together.
- The pandemic of HIV/AIDS: One of the salutary side effects the otherwise disastrous pandemic of HIV/AIDS is that it helps us to turn our attention once again to the values of love, marital fidelity, etc. which are the bedrock of the institution of family. Though AIDS spread is not only through sexual relations, it necessity be admitted that in a majority of cases sexual get in touch with is the main source of infection. Marital fidelity has, thus, the added advantage of also preventing a lot of infection. It is to be noted, that it is not HIV/AIDS pandemic that makes marital fidelity a value. It has been a value which was the foundation of the institution described family, and now in the context of HIV/AIDS we get the motivation to reflect on this significant value once again.

Society is the name for group of people who live together. A society is composed of several dissimilar kinds of organizations and institutions and family is one of them. In fact the family is the primary unit which constitutes society. Family, community and society are regarded as special dynamic institutions which interact in the middle of themselves.

It is the society which decides the excellence and kinds of family. It depends on the healthiness of the family to create a vibrant and progressive nation and society.

Again, family consists of individuals. So it is necessary to develop healthiness and proper understanding in the middle of each individual to make a family vibrant and progressive.

So the individual is the vital entity in a family. Family, community and society are dynamic in the sense that with the changes in outer world, it economics and history of these institutions are being constantly redefined and the boundaries are being constantly readjusted.

A healthy society or community is one large cohesive family. Family is a microcosm of society or community possessing all the culture and civilization it has acquired.

This can be best defined through the concentric diagram given below. This model shows that the individual is the cornerstone of the world community.

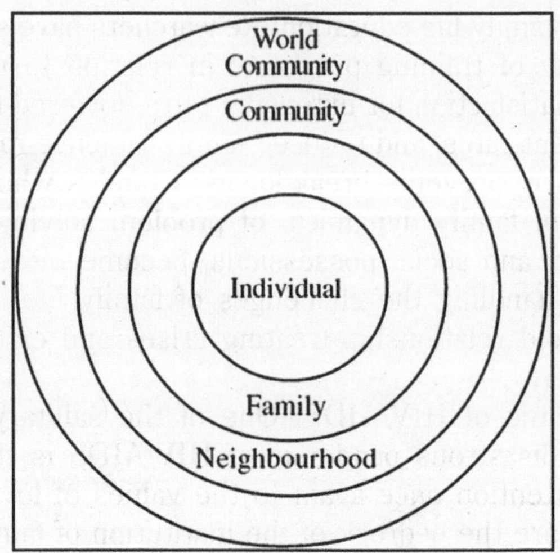

This is because s/he is the essential constituent part of family, neighborhood, community, society, nation and of world community at large. So any change in the individual's attitude and behaviour affects all these above institutions. Simultaneously, any change in the large institutions like neighborhood, community, nation and world community will affect the individual too. So it is pertinent that roles of all thcsc institutions arc crucial as far as family life education is concerned. Each institution has to play specific roles to stay harmony in the middle of themselves and in the world community.

ROLE OF INDIVIDUAL, FAMILY AND COMMUNITY IN FAMILY LIFE EDUCATION

Let us now turn our attention to understand the importance of an individual, family and of community from the family education point of view.

Individual

Family life education focuses on how individuals as constituent parts and fruits of family life may become physically healthy, emotionally mature, disciplined, responsible and tolerant human beings. Its goal is to help individuals develop interpersonal skills and more enriching human relationships through learning how to relate effectively within their families. As a whole it seeks to improve the excellence of their life throughout the whole range of human growths.

Human beings perform dissimilar roles (like that of husband, wife, parent-child) and functions (like child care and development, family resource management, family planning). That is why the same individual is supposed to know in relation to thrall these characteristics, their prospects and consequences. All these needs can be categorized under dissimilar headings. An individual may have several kinds of needs. Some needs may be the felt needs of an individual in a concrete situation in life. For instance, an individual may feel: "I need to be more assertive in my life relationships". The second kind of needs is "developmental needs" which are common to mainly individuals as they grow and get matured (for instance, dealing with one's changing sexuality at puberty, preparing for retirement as the age advances). These needs are usually recognized through empirical studies and through communal wisdom and experience of mankind.

In several cases there is an overlap flanked by these two kinds of needs. The educational needs of new parents, for instance, may be both felt need expressed through the parents themselves and a developmental need emerging from their new roles and responsibilities. There is another category of needs that deserves special attention which is described "societal needs". These are needs which emerge from current and anticipated social, economic and political circumstances that impinge in significant ways on the lives of all individuals. For instance, the pace of social change in mainly societies dictates a need to redefine the roles of adult family members. Where traditions once recognized the norms for family behaviour, parents are now being challenged to look at their role and to accommodate the demands of changing family structures and functions. The state also has shown an rising tendency to intervene in family life for reasons of economic and social necessity.

National programmes of Planned Parenthood, for instance, have been recognized in a number of countries to curtail population growth and to promote improved standards of living. The belief, that parenting skills, knowledge, and attitude can be learned through programmes of education as opposed to learning from one generation to another has given rise to a host of such programmes approximately the world.

Role of Family

Without the role of family, one cannot think of family life education. Individuals usually perform sure functions within the family which is not possible for them to perform outside the family. In other words, family promotes sure functions amongst individuals. These functions are discussed below in relation to their role in Family Life Education.

Biological Functions

The husband and wife have the right to sexual expression as a way of sharing or showing their love for each other. This provides not only sexual gratification for the spouses but also strengthens their interpersonal relation, love and unity as a result of the powerful and often binding emotions which accompany sexual activities a husband and wife. The sexual functions also help to stabilize the society. The family provides opportunities in which sexual drives find both expression as well as control. Parenting also has extensive term functions as a child is very helpless and dependent on its parents for food and shelter for a number of years.

Cultural Functions

Family helps the child to acquire values, beliefs, customs and traditions of the society. It is in the family that the child's vital attitudes develop. If puja is performed in a family in the morning, then children also learn and acquire this value of prayer and religiosity. What a child considers right or wrong mainly depends on what the family believes in and puts to practice. They may, later on in their lives, question these values and family traditions. It is in the context of this questioning that a child develops its own value system for life.

Economic Functions

There is a concept of optimum utilization of income and labour. Families pool their possessions so that they can have maximum satisfaction through budgeting efficiently. This economic co-operation within the family not only goes a extensive way towards fulfilling the economic function for society as a whole, but provides "rewarding experiences for the spouses working together which cement their union to work towards family goals such as providing good education, health and comforts to their children. Families sometimes decide to supplement family income through several ways such as wife's employment, business to uphold the social status of the family.

Emotional Psychological Functions

Human beings are emotionally and psychologically sensitive and need the family to recharge themselves for their thrash about in the world. In times of crisis particularly, the family works as a shock absorber, provides stability and anchors the members emotionally. The picture of family is one of inclusion and of caring.

One can quit his/ her job, drop out of school, and move to another city or village, but the family is always at the back to give support. Adolescents, children and unemployed youth are supported through the family so that they do not loose heart and go astray. The family acts as a compensatory agency which offers comfort and esteem to the worker who returns from the factory or office tired dissatisfied and without being able to exercise any control. With the changing family values and norms, members sometimes do not find the required support from it. Those who consider their own family as a "problem family" should learn to understand the root cause of the troubles in the family and to try to work on them to resolve them, through shared feelings.

Educational Function

Beginning with toilet training, language skills, or socially acceptable behaviour, a family performs the significant functions of education of children and their socialization. Vocational skills of varying nature are also learnt at times within the family. Vital education is given through the family. This training makes a child quick to learn in school and enables him/her to cope with the demands of schooling beside with his/her classmates. Within the family the child learns to conform and to submit to power. It also learns to negotiate with power.

Cosmic Function

Man feels adrift in the large impersonal universe. He becomes mainly alienated in the industrialized and urban society. Living in a family helps to provide him/her a place in the cosmos that makes him feel linked to others and rooted to earth.

Community

MacIver defines community as "an area of social living marked through some degree of social coherence". We know that a person rarely exists alone. S/he is connected in several ways to his fellow human being who form a group. S/he can establish relations only with the people who reside close to him/her in a definite part of the territory.

It is a fact that people who reside in a scrupulous locality in excess of a length of time, develop social likeness, have common social ideas, common traditions and a sense of belonging. This social interaction in a common specific area gives birth to community.

These above mentioned characteristics of community help in providing and preserving family life education. Let us now look at the three specific roles performed through community which are very significant from the family life education point of view.

Community acts as preserver of values related to family life

Each older generation passes on values related to family life acquired from their previous generation to each successive younger generation. That becomes a set pattern or tradition in the communities. This way, age old traditions and values still prevail in communities and societies. Sometimes we find that these values are little bit customized according to the changes taking place in societies. It may be either in the form of parents or grand parents vis-à-vis their respective children or grand children or in the form of sermons through religious or community leaders.

Community act as a resistant

Not only does community preserve values related to family life but it also resists when there is any violation of the norms or values. It imposes sure restrictions or takes action against the individual who violates the norms and values. Other than this, a community may take strict events like non-cooperation and boycotting the individual and groups who tend to violate the values and norms.

Community provides moral policing

Another important point is that a community always tries to give guidelines to the members concerning what to do and what not to do. It keeps an eye on each member's actions in the community or outside the community. In this way it keeps checks and balances in the community. As far as a community is concerned, there are three major settings currently which offer family life education for adults: religious, social and professional organizations, mass media; and schools and universities.

Role of religious, social and professional organizations

The mainly comprehensive programs in educating for family living have been urbanized and implemented through religious and secular organizations. Hundreds of diverse organizations and agencies are involved in offering a wide array of courses, workshops, and services covering areas such as parent education, marriage preparation, marriage enrichment and child development. Traditionally invested with a belief in marriage and the family, the church is one institution in western society which encompasses all families and whose clientele spans the whole life cycle from birth to death. Several religious groups are involved in matters relevant to the well-being of families. Besides religious denominations, there are community organizations such as the Young Men's Christians Association (YMCA) and Young Women's Christians Association (YWCA) and Service Association of America as well as voluntary organizations that contain a focus on family education in the form of courses and workshops offered to adults on issues related to marriage and family living.

Mainly countries have urbanized organizations to deal with their scrupulous concerns in family life education. These often grow out of professional associations having a special interest in the family or government agencies giving leadership to educational, social and health concerns of the family. Some examples contain the International Union of Family Organizations. (France), the National Council on Family Relations (USA), the Study commission on Family (UK) and the Vanier Institute of the Family (Canada) Separately from these organizations, several international agencies also sponsor this kind of work. These organizations are UNESCO, International Red Cross Organizations for Economic Co-operation and Development (OECD), and Cooperative for American Relief Everywhere (CARE)

Role of mass media in promoting community attitude

There continues to be a great public power exerted through television, radio, news paper and magazines on molding the values, ideas, attitudes and beliefs in relation to the marital and family living. While the excellence of input may be at times questioned, they nonetheless offer alternative structures, role models and information and advice on a wide range of family-related issues.

Role of schools and universities in providing family life education as a part of community education

Schools have for sometime been introducing more and more learning opportunities for students on marriage and family issues. Several universities and colleges offer credit and non -credit continuing education programmes on marriage and family for adults.

There is a need for an integrated effort or coherence flanked by family and community. If there is a variation flanked by what is observed in the family and what is practiced in the community, then there will be crisis in individual's mental stage or way of life. To avoid that there is need for orientation of community leaders. These leaders may be political leader from Panchayat stage up to the national stage, religious leaders of all religions and social leaders. These are the persons who plan and execute all the decisions which are relevant from the viewpoint of family life education.

ROLE OF HOME, SCHOOL AND RELIGION IN IMPARTING FAMILY LIFE EDUCATION

ROLE OF HOME IN IMPARTING FAMILY LIFE EDUCATION

Home is the ideal place to tell the story of life and to provide training in family life. This is because parents have a continuing relationship with their children in excess of several years. This relationship is crucial in developing attitudes as well as in sharing information. The way people act in adult life is strongly influenced through their experience and training in the early childhood at home. The vital attitudes recognized in the home throughout the early years will dominate later attitudes. So the family has to take up the major role in imparting education at the several stages of a child's development i.e. physical, emotional, intellectual, social and spiritual. These are the basics for successful family life.

Parents as Educators

Home is the school of schools and parents are teachers 'par excellence'. The child's intimacy with the parents throughout the formative years powers the personality of the child. Again the personality of an adult individual is usually influenced through the childhood experiences at home. John Locke, says that the child's mind is a 'Tabula Rasa' (Latin, meaning 'clean slate'). Parents can write anything on this clean slate. All the knowledge in relation to the family is derived from the experiences of the child's early life. A home provides every life situation in which a child needs training. It is the first socializing agency.

A family's standards and values act as a guide in later family life. A home also provides opportunities for open expression of personal, emotional and social behaviour.

Role of Home in Character Building

How does character affect family life? The success of a family depends upon the character of the main couple. Home has an significant role in training the character of its members. Success or failure, good or poor adjustment, happiness or unhappiness is much more a matter of character than of intelligence. Character impels a person to act on sure principles. A person of character can be trusted and relied upon. Hence character training at home is significant to lead a full life. In the modern world, lack of good leadership is felt in all walks of life.

Parent's necessity assumes the responsibility of assisting their children to become socially competent adolescents, mutually understanding spouse parents and honorable citizens. Happy homes tend to build happy people. If the parents are leading a harmonious life, children also can build similar marital life relationships. A child who fells the love, security and sense of belonging, and who was cared for in a secure happy home, makes a good start towards a responsible and matured family life.

Education for Marriage at Home

Who should start imparting education for marriage? As you have seen, home is the starting point for all education. A child's image of his role as a boy or girl, husband or wife, father or mother is patterned very early in life through the emotional and social atmosphere enjoyed at home. What does a child learn learn from his parents in relation to the marriage and family? From parents the children learn the good or bad of family life. They learn the art of loving and being loved, giving and getting affection, adjusting and sacrificing from the family. The first lesson on the art of human relations and adjustments are learned from the family. The girls learn the art of home making and boys the masculine skills, from home. Parents have the moral responsibility for preparing children for married life. Family life education at home inspires the child to understand the nature and meaning of the home itself of which he/she is a part. It prepares him/her for his/her own future family life.

HOME AND PERSONALITY BUILDING

What is the role of home in building one's personality? What a child becomes is more significant than what a child knows. A good personality is the mainly valuable gift that parents can provide their children. For normal development of personality, child should have a satisfactory family life. Homes in which parents and children share their joys and troubles, where they participate in recreational activities with friends and relatives, where there is always peace and happiness, will produce well adjusted and socially confident individuals. These children will have personal adequacy and personal responsibility in their relationship with others.

Family living provides sure vital human needs, more directly, than is possible elsewhere. Parents who really love each other, provide a sense of security to their children. These parents provide an unconscious model of human relationships. Children will understand that inspite of disagreements and quarrels, the vital loyalty of the family group is never threatened. They will be able to establish strong family relationships later. This experience can never be given through books or study classes. It comes only through the day-to-day life in the family. From where does one inherit habits and behaviour? Certainly inherits mainly of one's good or bad tendencies from one's parents. These tendencies take the shape of habits or behaviour through interacting with the environment, in which one grow. The factors significant in personality development are heredity, environment and training. All these factors are contributed through home. The way the twig is bent, so inclines the tree. A thousand mile walk starts with the first step.

Role of Home in Sex Education

Are sex education and family life education identical? Even though they are mutually related, sex education cannot be equated with family life education. Of course, sex has a vital role to play in family life. Today, researches show that intimacy, love and companionship are the mainly significant needs of couples. These needs are satisfied mostly from sex relations. This elevates sex to a position of new importance in marriage. According to several people, the success or failure of marriage depends upon satisfactory sex relations. This emphasizes the importance of sex education for a successful family life.

Unfortunately children are exposed to "too much too soon" and with little moral direction and guidance. So home has a major role to provide this direction and guidance in sex. But since there is a separate section on sex education, this unit is not taking up a detailed study of that topic.

Difficulties in Exercising the Role Expectations of Home

Do all homes have the facilities and competence to give training for family life? Unfortunately no. If so, what are the blocks?

- Failure of parents: Several families today have only one or two children. So parents do not permit their children to encounter any hardships or hazards or ordinary risks of life. There is a tendency of in excess of – protection and in excess of-ambition. The in excess of – protected children can not face the normal risks of life in their later family life. They have a very sheltered subsistence.
- In excess of-expectations of parents: Parents expect too much from children. The education system is also highly competitive. Children who cannot come up to parental expectations may develop several behavioral troubles. They may not be good partners in family life.

- Rejected children: Parental rejection is another problem. A parent who neglects a child rejects it. Now that a good majority of women are also working, children have two fathers – two providers – and no mother at home. A working mother who prefers office work to child care is a rejecting parent. Parents who do not provide their children attention, time, affection, companionship, supervision, discipline, religion etc. are depriving them of a cozy and secure home. This rejection results in attention seeking behavior, hostility to family member's frustrations, anxieties or extreme independence. These children may not be able to become good partners and parents in their own family lives.
- Perfectionist parents: Some parents have unrealistic fantasies of model, spotless, scratch less, and perfectly regimented homes. Their children may become very formal rigid and unrelated. They may develop distorted personalities which may end in marital breakdowns later.
- The indulgent and domineering parents: An indulgent parent gives excessive privileges to children. This results in immaturity which later leads to drugs, alcoholism and an irresponsible way of living. Domineering parents make all decisions for their children. Also they never allow children to grow to maturity. These children go through life with fear and a disrespect for all power. Children of both indulgent and domineering parents develop distorted personalities and distorted ideas in relation to the marriage and family.

Traits in Parents for Success

What are the qualities that help a parent in bringing up children? Being a parent today is more than challenging: it is a full time commitment. In reality, the job of a parent is demanding and hard, but at the same time rewarding and fun. Child rearing is a responsibility and not a sacrifice. Enjoy parenthood and work for it. Be aware of the limitations. What is caught from the parents is more significant than what is taught. Parent's necessity meets the varying needs of children at dissimilar ages and stages of growth. They should give a sense of emotional security and inculcate self-reliance and self-confidence in children. Each child is endowed with some innate potentialities. Parents should give time and space for the free development of these innate abilities. In excess of-supervision is not good. There should be discipline – firm-but not harsh. Understand the child as he/she is from his/her viewpoint. Develop a sense of responsibility in children. Proper communication is the key to settle mainly of the troubles flanked by parents and children. Through understanding the child, you can avoid the bitterness of anger and confusion. This restores the live scared through you and your child.

The parents have a role in imparting family life education. "A river can not rise above its source, so also those who teach values cannot do so unless they themselves hold these high" Today, the structure and function of home has changed a lot. Several of the functions of home are taken in excess of other external agencies. Here comes the importance of school, peer group and religion in family life education.

ROLE OF SCHOOL IN IMPARTING FAMILY LIFE EDUCATION

Can the school do anything to impart family life education? Several people underestimate the role of the school. Even though the family has the primary responsibility for family life education, the school also has an significant role. School is an extension of the family. Since the school has a regular and continuous get in touch with all children, it has a unique opportunity to supplement family training. School age is the period at which a child's world widens slowly, He/she comes into get in touch with other socializing agents such as teachers, other adults and peers. So school can give more opportunities to learn in relation to the family life.

Why should the school interfere? Several parents do not have the technical knowledge to answer all the questions their children may ask in relation to these and family. They are not able to help children to develop healthy attitudes and to make correct decisions and choices as they grow up. The informal education given through parents often fails to achieve the goals of family life education. In several Indian homes the parents are illiterate and not competent enough to provide training in family life. Several of the old values and norms of family life are changing. In these circumstances, the school is a more reliable agent than the home to impart family life education. The school can provide more detailed and systematic teaching than the parents.

Is their any barrier to family life education at school? Usually family life education is equated with sex education courses. Several parents are frightened or suspicious of group teaching in this area. Of course, sex education has an significant place in these courses, but it is not the only-not even the primary purpose of a course on family life. Parents should be conscientised in relation to this. A vital goal of education is to give a background for choices in life. To make intelligent decisions in the area of family living, children should get correct information in relation to these and family life. A co-operative effort is needed flanked by the parents and teachers in the matter of family life education. In school, the information can be imparted in a much more impersonal way without anxiety and tension.

Role of Teacher in Imparting Family Life Education

In school a teacher is the key person in imparting family life education to children. The teacher should be competent, gifted and a prudent person. Every teacher in school can teach family living. Every subject and every action at school can contribute to a better understanding of life. Experiences in school enhance the formation of ideals and values. This in turn helps to mould the character of children. It also leads to personal and social adjustments. As said earlier, sex education is one of the mainly significant characteristics of family life education. A teacher can provide this education without much inhibition can impart a noble understanding to this deepest personal mystery. Students tend to approach their teachers with personal and general troubles. Children prefer several issues to talk about with teachers than with parents. They believe that teachers are better informed and better adjusted persons.

Teachers help children to understand themselves better, and to make better adjustments in life. A teacher's task is to impart information, clarify doubts, and provide guidance – in short to be a mentor. Armed with knowledge, rather than ignorance, the young generation can face the challenges of rising up and has a better understanding of the values of sex and marriage as well as its pitfalls.

The Role of the Peer Group

What is the contribution of the peer group to family life education? Once they begin to attend school, children spend mainly of their time outside home, with members of the peer group. So peers have a greater power on the attitudes, interests and behaviour of a child than the family has. The peer group helps the youth to get socialized, conforming to social patterns and norms. The youth is more open and free with peer group. Again the peer group serves as a laboratory for developing variety of physical and social skills and social roles. It plays the role of teacher in this procedure. It provides a sense of security and normal backing when the adolescent wants to be self-governing.

Inspite of the powers of the peer group for better adjustment, there are sure risks involved in the experience and information gained from the peer group, especially concerning sex and marriage. The information gathered may be wrong and dangerous. The peer group members may be ill-motivated and inexperienced persons. Such inaccurate and inadequate and even distorted information in relation to these and marriage creates unhealthy and wrong attitudes towards sex and family life. It may also lead to serious mental disorders, complexes, and maladjustments in behaviour. Several troubles in married life like impotency, frigidity, guilt feelings, sex perversions, excessive sex interest are the result of some of the negative power of the peer group.

ROLE OF RELIGION IN IMPARTING FAMILY LIFE EDUCATION

Family life education was given through the families, in an informal way through all religions. In the Indian context, religions occupy an significant place in the life of an individual. Marriage and family are measured sacred and hold through all religions. Each religion has its own laws and regulations for marriage and family. The culture of a family in dissimilar religions is dissimilar. A strong religious background is necessary for a successful family life. Today it has become a fashion in the middle of youth to disown God and provide importance only to the material world. Secularism is a wrongly understood word for the youth. According to Sri. C Rajagopalachari, secularism does not mean that boys and girls will be deprived of the advantage of the disciplines based on the religion professed through the families to which they belong.

So far, a formal training for family life was not seriously thought of through religions. But due to the high percentage of marital breakdown and troubles in family life, all religions are now taking up the matter seriously. Several religions, especially Christians have come out with some positive steps in family life education. Attempts are made to provide orientation in family life. Premarital and post marital courses are offered. Family counseling facilities are organized. Church centered programmes are gaining momentum and getting wide acceptance. Several couples come to these centers for help and guidance. All religions have the responsibility to give family life education for its members. The youth should be given opportunities to learn in relation to the marriage and family before they enter into them. Religious values can reinforce the personality of an individual.

METHODS OF IMPARTING FAMILY LIFE EDUCATION

Family life education is a life extensive procedure. It comprises the knowledge, skills, attitudes and values involved in maturing into men and women. This education should help the individual to find his/her personal role as a member of one's own sex in marriage, family life and the community. How can we provide this training? What are the methods to be adopted in imparting this education? Marriage and family occupy a very significant place in the life of a person. But very few couples have any experience in the job which requires special knowledge and training. The importance of this training is not properly understood. Usually, parents are expected to learn on the job. This may lead to several mistakes. Mainly parents rely on what they received from their elders on their own upbringing. They also depend upon the advice from their parents and friends.

Often commonsense knowledge and child care books give the guidelines for young couples. This is not good enough. For any other job, specified for instance, teaching, nursing, driving or any other profession, specialized special training is necessary before one enters the profession. In-service training is also given for mainly of the professions. But for mainly people pre-marriage training is not measured as a pre-requisite for marriage.. It is not easily accessible also. But today the importance of this training is widely accepted. Since family life education is mainly given through the home and school, we shall talk about the methods adopted through them at dissimilar stages of development-pre-school stage, primary stage (age-6-9) pre adolescent stage and adolescent stage. Let us consider them one through one.

Pre-school Stage

This is the stage where the whole responsibility of family life education is on the family. The school does not come into the picture. The vital attitude recognized in the home throughout early years will dominate in the later attitudes. Children spend the early formative period with parents and other family members. So the initiative in family life education should come from home. Throughout infancy and early childhood, mothers have to take the lead. The vital pattern of the personality is shaped throughout childhood. It is said that even 'sex education' is given through parents to their young child, several years before they can begin to provide him or her sex information.

To a rising girl, the mother is the role model. A little girl will try to identify with her mother in several ways. But the girl should feel the warmth of her father, his voice, his touch which is dissimilar from that of her mother. The father's strength and confidence, his behaviour towards the daughter stamp the image of an ideal man in her young mind. If she misses this power, it will be hard for her to understand and appreciate boys and men in later life. This experience of man – her father – will stand in good stead in adjusting and understanding her husband when she gets married. A rising boy learns from his father and mother. The mother's concern, affection and care for little things tenderness, warmth, patience, all provide him the image of an ideal woman and that is what he is looking for in this wife. The father shows him that a man can be firm and strong, yet kind and considerate.

It is not easy to perform the special tasks of a home. But the attitude of the parents and the spirit in which the household duties are done, will shape the child's attitude towards masculine or feminine roles later in life. Pre-scholars are always curious. For them everything approximately them are new and exciting. According to Freud, the first six years are the formative years in the life of a person. Hence the experience received at this stage is very significant. We have already discussed the informal methods adopted through parents in training their children. It is usually an informal and indirect procedure. Education comprises not only what is taught, but mostly what is caught through an individual from his/her everyday experiences in the family.

The child learns from the examples set through the parents and elders. The physical and emotional environment of home is very significant for harmonious, peaceful and serene living. The mainly crucial power an a child's development is the training given at home pertaining to orderliness, system, punctuality, cleanliness, books, art, music, prayer. These are vital factors in building up a family. Learning through doing should be the method adopted for giving training in family life. Provide the child enough opportunities to participate in all the activities of the home related to work, recreation, social and religious activities. From day to day experiences of family life, children slowly absorb the strength of character, intelligence, emotional and physical maturity, religious and moral values of their parents.

Should there be a formal education for family life?

In addition to the informal and indirect methods of teaching, parent's necessity is prepared to answer the questions asked through the pre-schooled. Parent's necessity is careful in providing answers to a Childs, queries. One should not resort to fairy tales to answer the questions. Intelligent answers to their stage of maturity and comprehension necessity are given. It is better to use scientific terms as far as possible. But unfortunately, in Indian languages we do not have correct scientific terms. Before, answering the question, stage of a child's knowledge should be taken into consideration. A parent should be a good listener too. Frankness, honesty, uniformly. If the parent refuses to answer, children feel inhibited to ask further questions.

School years 6-9

How is the school age child dissimilar from the preschooler? At this age a child becomes aware of gender roles. Appropriate role models become necessary too. At this stage, also the mainly significant guidance a child can get from the way the parents act as adult human beings. The attitude of the parents towards each other's work and duties will be reflected in through children's. thinking. If the father looks down upon the mother's work, or thinks that his is the 'provider and head' of the family, children will develop similar attitudes later on. The mother's attitude as a woman, wife and mother, her acceptance of her own family, and the father's attitude to his masculine role, his firmness, his sense of purpose are predominant factors in the formation of the personality of a child. The tenderness and warmth flanked by the parents, the special bond flanked by father and mother, and physical affection become good experiences for the child. It makes children understand that affection is permissible, desirable and significant. When they later learn the facts of life they know it is an expression of love.

What in relation to the questions asked through children?

Children ask too several questions in relation to these and family life throughout this period. Never assume that 'they know'. This is the proper time to teach them in relation to the menstruation and puberty. There are two things to be remembered in giving sex education to children at this stage. It should not be "too little and too late" and "too much and too soon". Don't get embarrassed to answer the questions. The answers given to the questions at this stage add to what has been gained earlier. This will be the vital preparation for the time when sex begins to concern the child more personally. Provide children confidence and encouragement. Repetition does not hurt. It provides more opportunities for more questions. Allow them to play with their friends and get socialized. But parents should always stay 'an ear and eye' on the children at this age.

What is the role of the mass media?

Mass media like T.V, radio, cinema, books etc. help to widen the child's knowledge in relation to these and family at this stage. Whatever the source the knowledge and value they gain will be part of their foundation for life.

What can the school do?

In school, the curriculum and methods of teaching should be suited to the stage of comprehension and maturity of children. Sex information, which is a part of family life education, can be imparted through biology lessons. Children can be taught in relation to the concept of maleness and femaleness, reproduction in plants and animals, the ideas of fertilization etc. There is no need to teach in relation to the human reproduction at this stage.

Pre-adolescent stage

This is the best stage for acquiring vital knowledge on family life. A rising child wants to learn, to know. He/she observes everything that happens in the social setting of home, school and neighborhood. At this age he/she tries to understand and to work out the meaning of what he/she observes. So the home and school should give ample experiences for the child to understand life approximately him/her. As the child grows up, help him/her to feel proud and happy to be a male/female. In the extensive procedure of rising up to adulthood, children may have to face several challenges in life at one time or other. Imparting love at home is absolutely necessary for the development of a child. A home is a place where the child is needed, wanted, appreciated – where it is good to be. But love should be expressed. Love flanked by the parents and flanked by the parents and children is the very essence of good family life education. Loving his/her father and mother and other members of the family, the child later learns to love his/her partner in marriage, and his/ her own children.

How can parents cater to good fatherhood and motherhood?

Parental power is the mainly significant facto in molding a child's attitude towards fatherhood and motherhood. The examples the parents set throughout the years of their children's growth is the criterion in deciding whether the children will have a happy or unhappy family life. Children should see the complete, dedicated relationship flanked by their father and mother. Parents themselves will have to grow in order to be wise parents. If parents want to do their best for the children they necessity begin with themselves.

How does the peer group power the preadolescent?

This is the time of peer group power. Parents should be aware of their children/s friends. If the friends have attitudes and values that are dissimilar from their own values, they should talk to their children in relation to it. Try to guide them towards friends who possess good moral values. There is no harm in getting attracted to the opposite sex, but there should not be any sexual misconduct.

Should there be any limit to mingle with the peer group?

Parents have to set limits for their children and to guide them. Talk in a rational and loving manner. While discussing these matters with children, be courageous and firm. Avoid a sex lecture. The communication should be light and friendly, but always honest. Provide information in relation to the child's own sex as well as the opposite sex. Teach the child to respect the opposite sex.

What is the power of the mass media?

The environment in which the present day child grows has changed radically. Sexual matters are projected everywhere through dissimilar mass media. The living atmosphere is saturated with sexual awareness. No one can be shielded from its impact. Issues like rape, abortion, teenage pregnancies, S.T.D, homosexuality appears regularly in the mass media. This period is the proper time to instruct children in relation to the these. At home the father can take the responsibility of educating the male children and the mother can reveal the facts of life to her daughter. Grandparents and other elders at home prove quite helpful in this task.

What should the school do?

After the parents, the school should be taking up the main responsibility of educating the pre-adolescent in family life. The school can provide more detailed and systematic teaching than the parents. Appropriate curriculum should be drafted to cater to the needs of youth.

What kind of sex education should be given at this stage at school stage?

There is a strong need for sex education at this stage. It should equip one with correct scientific information to face the challenging situations of adolescence. Before, one attains puberty, sex education can be imparted without involving emotional reactions. The education preferably should be given separately to girls and to boys. Classes of general hygiene, physiology and programmes related to physical education and personality guidance can be utilized for this purpose. There necessity is standard literature related to sex in school libraries. All the media can be used positively for this purpose.

Adolescence

What is the special importance of this stage? This stage has a tremendous impact on the rising youngsters. It is a transitional period in personality development bridging the years flanked by childhood and adulthood. Like any other period of transition, adolescence brings a number of adjustment troubles. Adolescent's needs, interests and troubles are dissimilar from those of childhood and adulthood.

Who can help the adolescent?

Parents, teachers and other elders should recognize if adolescents are facing any troubles. Parents have a vital role to play in preparing the adolescent for his future family life. At the home the adolescent learns what it means to be a man or to be a woman. Through his family experience the boy is forming the concept of man, husband and father. The girl learns the role of woman, wife and mother. Parents have to accept that the child is rising to be an adult. Parent's necessity varies their protective role and provides more initiative and freedom to the youngsters. Mothers have a tendency to prolong the dependence of the child. This will not help them to grow into adulthood.

You may wonder why the interests of the adolescent move from his/her parental home. This is only because up to this stage the child has everything in his/her home circle to meet his needs. But with adolescence, sexual needs emerge and their interests move outside the family circle. Life steers him/her absent from his parental home towards a home of his/her own. He/she wants to establish a home of his/her own with a partner of his/her own choice. Parents need not grow anxious or unhappy in relation to the youngsters' need for getting married and founding his/her own family. Special efforts necessity is made through parents to train and help the adolescent for choosing his/her partner and establishing a family.

Should the adolescent learn through trial and error?

Some parents think that they need not interfere with the troubles of their young boys and girls. This should not be so. Of course, the role of the adult is not to protect him/her from all adversity or every painful experience. Parent's necessity let the adolescent do things for themselves. Then only they will become responsible adults. The role of the parent is only to guide and assist the young person when he/she has a real problem. They should enjoy the procedure of a child rising into a young adult. Adolescents need love and reassurance, recognition and praise from the parents to develop self-confidence, belief and trust in themselves and to accept themselves as they are. At home he/she experiences the way in which the two sexes complement each other, feminine gentleness and masculine firmness. It is not formal learning, but he/ she slowly acquires the philosophy of living in a two sex world, through living in a family. The attitudes of the parents will power his/her relationship in marriage and family. It will make him/ her a self-reliant man or a women, a confident husbands or a wife.

How can the school help the adolescent?

At this stage the school can provide a very systematic programme of family life education. The educators and the people in power should plan for a curriculum which comprises the essential information and learning experience in family life. This will provide a formal recognition to the importance of preparation and training for family life. To make intelligent decisions in the area of family living, children, should get correct information. They should also develop proper attitudes. Knowledge gives them a feeling of security and protection. This in turn helps them to develop their potential to live in a responsible and a constructive way.

Sex education is very significant at this stage. Sex is a not a "subject" like Arithmetic or History. Hence it is not advisable to offer it as a subject. It is vitally significant that young people should receive honest and accurate answers to their questions concerning love and sex. But they should be given in a simple way.

What should be taught?

Human reproductory system can be taught at this stage. Children can be introduced to concepts relating to sexual maturity, conception, childbirth, illicit sex relations, STD, AIDS etc. There should be healthy hetro-sexual relationship in the middle of boys and girls. The content should be functional– that is directed at the specific troubles young people face in their adjustments to the opposite sex throughout the pre-marital period and in early years of marriage. But unfortunately we do not have sufficient material in this area to give adequate guidance to the youth. Though, block 3 of this course deals with mainly of the matters pertaining to the physiological components of rising up. Family life education, at this stage should give opportunities for the youth to observe nursery schools in operation, observe the care of small babies and learn more in relation to the parenthood and child training.

Discussions and seminars on family life education can be arranged. Topics like responsible partnership, parenthood, family planning, family relationships, psychology of man and woman, how to choose the right partner can be incorporated in such discussions and seminars. Help from persons like psychologists, medical doctors and other specialists who have first-hand knowledge in relation to the family life can be taken. There necessity is trustworthy possessions accessible to the youth. The educator's necessity is well-informed in the subject. Counseling facilities should be provided in the school to help children clarify their doubts and personal troubles. Pre-marital courses can be organized. All the mass media can be utilized for this purpose. Good books on family life can be provided in the library.

Is there any other agency to impart family life education to youth?

Social and religious organizations can play a vital role in giving family life education to the youth. Lately, several social and religious organizations have been arranging life orientation programme for the youth. Today teenagers, newly weds and extensive married couples are approaching such agencies for guidance. Family courts and family counseling centers are also doing a very good job in helping young couples and youth in family matters. Some newspapers and periodicals publish regular columns to address issues pertaining to the common concerns of people on sex and sexuality.

DEVELOPMENT OF PERSONALITY AND MORAL VALUES IN LIFE

Nature, Definition and Development of Personality

You know that personality is of great significance in every walk of our life. Naturally you may ask sure questions. What is the nature of this highly valued thing? Do all of us have personality? Is personality determined at the time of birth? Why are some people are calm and quiet while some others are hot tempered and aggressive ? In this chapter, we try to find out answers to these questions.

The term 'personality' is derived from the Latin word 'Persona' which means 'mask'? In ancient Rome, actors used masks to hide their identity on the stage. Hence persona denotes 'as one appears to others, not as one actually is'. Persona does not indicate the internal organization of an individual which affects external behaviour. So persona does not represent the real personality.

Personality is something deeper than the outward appearance. It describes a person's character, emotional disposition, sociability and others characteristics of what a person says and does. It is the product of a extensive procedure of his physical, mental, emotional and moral development. For the psychologist, personality is a dynamic concept describing the growth and development of the psychological system of an individual.

Definitions

It is very hard to find an ideal definition for personality. The first problem faced through us in the study of personality is the need to find out an ideal definition for this phenomenon. Dissimilar psychologists have dissimilar views in relation to the personality. They have defined personality according to their viewpoints.

We can go through some of the definitions. Munn has defined personality as the feature integration of an individual's structure, manners of behaviour, interests, attitudes, capacities, abilities and aptitudes. The scrupulous way in which these qualities are integrated in a scrupulous person determines his personality.

The mainly regularly used definition for personality is given through Gorden Allport. According to him, "Personality is the dynamic organization within the individual of those psychophysical systems that determine his unique adjustment to his environment."

This definition emphasizes the dynamic nature, i.e. the changing values of personality. It recognizes the importance of adjustment to the environment. It also stresses the importance of psychophysical systems, i.e., the habits, sentiments, emotions and motives of an individual which are psychological in nature, but have a physical base. The disadvantage of this definition is that it stresses only the theoretical part of personality.

How does growth occur? It occurs in two patterns enlargement and change. Take the examples of a leaf and butterfly. The leaf expands in size in the procedure of growth.

But it retains its essential form. On the other hand the butterfly passes through dissimilar stages of change to attain its adult form. Human beings manifest both kinds of development. As a child grows older, his/ her body becomes superior and heavier.

But like the leaf, the same structure is retained. On the other hand, the psychological procedure undergoes several changes throughout the early years of his/her life.

Personality development implies tracing the connection flanked by the personality of a child and that of an adult. It is the understanding of how adult behaviour and psychological processes evolve from a child. In other words how does the butterfly emerge from the caterpillar?

Need for Studying Personality Development

Why do we study the development of personality? There are several reasons for this. The mainly significant ones are the following:

- Studying personality helps us to understand the functioning of a personality in its mature form
- It helps us to predict adult personality and personality disturbances.
- We can understand the behaviour of a person only if we know in relation to the specific organization of his motives, beliefs, skills, attitudes etc. at dissimilar ages.

Molding of Personality

What makes personalities and what are they? There are no specific causes leading to changes in personality. But a number of factors interact with one another in the formation of personality and in its development.

- Heredity: How significant is heredity in shaping the personality of an individual? Heredity plays an significant part in determining one's personality. An individual gets his whole hereditary endowment at the time of conception from his/her father and mother through the genes situated in the chromosomes. The personality pattern of a person is actually framed at this time. A child inherits the main raw materials of his/her personality – physique, sex, intelligence etc. Later these develop further on his/her interaction with environment.
- Environment: Unfavorable circumstances in the prenatal as well as postnatal environment will have lasting and damaging powers on personality development at a later stage. The significant environmental factors that exert pressures on our personality formation are culture in which we are rising up, our family, friends social and group's community.
- Culture: Culture establishes the norms, attitudes and values that are transmitted from one generation to the other
- The family: A child spends his formative years in the family with his parents and siblings. The family is a small group. But the intimate relationship in the middle of the family members powers the personality development of the individual in a powerful manner. The home environment, the school the child attends, his friends, the number of siblings, their personalities, the ordinal position of the child all contribute to his/her personality development. The attendance of harmonious, loving, peaceful family atmosphere foster healthy personality development.
- Economic environment: Unfavorable economic circumstances usually lead to the development of undesirable personality traits like inferiority feelings, nervousness, emotionality, lack of social initiative or sometimes even bitterness against society. Of course there are exemptions. But self made men/women are unusual.
- Social role: The social circumstances in the family are also decisive factors in personality development of an individual. Family is instrumental in providing training to a child to adjust to other people, to control his/her emotional expressions, and to cooperate in several fields of activities. There is a continuous learning procedure flanked by a child, its family, its social and community group.
- Situation: A third factor which has effects of heredity and environment on personality is situation. An individual's personality may be usually

stable and constant. But it undergoes changes in dissimilar situations. Dissimilar demands made in dissimilar situations bring in relation to the different changes in personality.

Relative Importance of Heredity and Environment

At this stage an significant question arises as to what plays a more significant role in personality development? Heredity or environment? It is not easy to answer this question.

Personality appears to be the result of both powers. Inherited qualities set the parameters or outer limits in shaping the personality. Social environmental factors can strengthen or weaken the inherited abilities. The social environment reinforces the development of inherited personality traits as the child grows older.

Heredity and environment are supplementary to each other. Personality development is not possible in the absence of either heredity or environmental factors.

How constant is an individual in his personality traits as he grows up? Studies show that distinctive qualities noted early in life tend to persist. It is assumed that there is a strong thread of consistency running through the life of an individual.

Theoretical Approach to Personality Development

Development of personality is a unique characteristic of an individual. Dissimilar psychologists approach personality development from dissimilar view points or perspectives. Can the personality theories help us to explain and predict the personalities of people? We do not get a complete picture of personality development from any one theory alone. Let us see what each theory has to offer. The best from each will help us to understand personality development in a better manner.

Biological Perspective

The biological foundation of personality is provided through researches mannered in this field. It has been found that 50 per cent of personality features are genetically determined.

The rest is determined through interaction with the physical and social environment. According to the theories of Darwin the inherited qualities help an individual to adjust to his environment.

Psychoanalytic Theory of Personality Development

This theory assumes that the vital dynamics of personality comprises the disagreement flanked by two opposing forces – anxiety arising from the inhibition of desires and defense against those desires that arouse anxiety.

Freud's Theory

The major contribution in this field comes from Sigmund Freud. He explained 'personality' as the subsistence of conscious, preconscious and unconscious in an individual.

Id, Ego and Superego

Freud also believed that our personality is built approximately three interacting processes, the id, the ego and the super ego Id is the unconscious reservoir of psychic energy. The other two systems operate on this psychic energy. There are two vital instincts in man – the 'life' or sex instinct and the 'death' or aggressive instinct. All the drive required through a man is derived from these two instincts.

Id is present at the time of birth. It does not know any morality or reality. It is concerned only with satisfaction of its wishes. It is guided through the pleasure principle. Ego functions as the manager of personality. It obeys the principles of reality. It controls, selects and decides what instincts of the id are to be satisfied and in what manner it can be done.

Super ego works on morality principles. It represents the values of society. Human personality is the result of the interactions in the middle of these systems. What will be the outcome if one of the systems dominates in excess of the other two? The attendance of a powerful ego ensures a well balanced personality. The dominance of super ego leads to a neurotic personality while supremacy of id in excess of ego and super ego ends up in the development of a delinquent personality.

Psychosexual Stages in Personality Development

In addition to the concepts of id, ego and super ego, Freud accentuated the importance of sex in personality development. He regarded the psychic energy for sex drive as the foundation of personality. So Freud explained personality development on the foundation of the five stages of psychosexual development.

- Erogenous Zones. In each stage of development a child derives pleasure through stimulating a scrupulous area of the body. These are recognized as erogenous zones. The environment, though, imposes restrictions in the satisfaction of his desires. How the child manages this disagreement is crucial in his personality development.
- Fixation. Too much or too little satisfaction in any stage results in fixation. This means that personality becomes emotionally fixed at a scrupulous stage. This is also detrimental to personality development.

The development takes place in five stages:

- Oral Stage (1st year): Pleasure is obtained through the stimulation of the mouth. Satisfaction at this stage lays the foundation for a man's adult personality traits like adjustability and tenacity. What are the traits urbanized if libido is fixed at this stage? Fixation results in the formation of passive personality. It is associated with in excess of eating, smoking, drinking, or sarcastic criticism of other's ideas.

- Anal Stage (2 to 3 years): Eliminatory procedure provides the focus of pleasure. The child achieves bowel control at this stage. We can see the beginning of the development of ego in the child. If parents are too strict in teaching toilet habits, the child develops anxiety. He manifests his anxiety and anger through excreting at the mainly in appropriate time and place. These are the prototype of all kinds of expulsive traits – cruelty, destructiveness, temper tantrums, disorderliness etc. On the other hand, praise for excreting at the proper time and place makes him feel that it is an significant action. This thought lays the foundation of creativity and productivity. What kind of personality is urbanized due to anal fixation? A stubborn, compulsively orderly personality is urbanized.
- Phallic Stage (4-6 years): This is a very significant period in the personality development of a person. At this stage pleasurable sensation comes from self manipulation of genital organs. In some cases sexuality produces guilt feelings in an individual. In some other cases the adult person tries to reduce anxiety through engaging in sex.

Oedipus Disagreement

This stage is also marked through the emergence of Oedipus disagreement in children. This is a intricate in which there is a strong attraction for the parent of the opposite sex and envy for the same sex parent. At the same time the child knows that it is a wrong thing.

He does not want to lose the love and affection of the same sex parent. To resolve this disagreement, the child tries to identify with the same sex parent. He/She tries to incorporate the sexual orientation, mannerisms and values of the same sex parent.

Resolution of Oedipus intricate results in the formation of super ego. Unresolved Oedipus intricate results in several troubles in later marital life. Such children can never get closer to their partners when they grow up nor make normal sexual relationships. A highly seductive female continues to feel guilty in relation to these.

Freud assumed that every person is bisexual. What does this mean? There is attraction towards members of the opposite sex as well as those of the same sex. This is the constitutional foundation for homosexuality. In mainly people this instinct remnants latent.

Latency Period (6-12 years)

Freud thought that sexual urges were dormant at this state. The fact though, is that this is the stage in which a person learns to befriend people of the same sex. The friends of a boy at this stage are other boys; those of a girl are other girls. A sexually mature person should know to be at home in the company of people of the same sex and of the other sex. This is a stage in which this learning takes place; The child also begins to achieve emotional independence. An augment in the knowledge in relation to thesis/her environment enhances his/ her ego development.

Genital Stage

This is the final stage of development. Sexual interest reawakens at this stage. There is interest in the opposite sex. Sexual attraction, socialization and planning for a vocation, marriage and family life begins at this stage. If a person is well adjusted in the previous stages, he will be capable of establishing normal hetro-sexual relationships. Mainly of the sexual troubles in adult life come from failure at earlier stages. There are no sharp rows dividing the dissimilar stages of development. The final personality attained through an individual comprises contribution from all stages.

Criticisms of this Theory

- Too much importance is given sex in explaining human behaviour.
- Freud's views are derived from his clinical experiences of people with disturbed personalities.
- Several points raised in this theory did not have empirical support.

Then what does Freud's theory have to offer? We cannot accept or reject Freud's theory as a whole. We can accept some parts, reject some others and revise a few. Freud put forward some challenging ideas which have relevance even today.

Application to Control HIV/AIDS

From the above theories we know that id works on the pleasure principle. The impulses of id are controlled through ego and super ego. But if id is more powerful it will not provide an importance to the voice of ego or super ego. It will not provide any concern for moral values also. It goes on enjoying pleasure giving activities. Through controlling the id, undesirable sexual life can be prevented.

Social Cognitive Theories

You have already learnt that psychoanalytic theory focuses mainly on the unconscious for the development of personality. Now, we shall go through the social cognitive theories. Psychoanalytic theory is a broad term including behavioral and social learning theories as well as cognitive psychology. Learning theorists focused their attention on observable behaviour. They also stressed the role of environment in personality development. How is the child's personality urbanized? The pioneers in this field, John B. Watson and B.F Skinner believed in a single Stimulus Response relationship in the formation of behaviour. They accentuated the importance of reinforcement and excluded the role of consciousness in behaviour. Mainly of the learning theorists followed the pioneers in this field. Though, in the modern times, behaviorism is concentrating on real life troubles with respect to human behaviour.

Observational Learning

Albert Bandura gave a new approach to personality development on the foundation of social learning theory. Bandura and his co-workers (1986) demonstrated that we acquire mainly of our personality traits through observational learning with or without direct reinforcement. A child for instance, observes and imitates the behaviour of a role model. The reinforcement received through the model serves as reinforcement for the child also. Like this, through observing such models, the child acquires several traits. This paves the way for the formation of his unique personality. The several component of learning (environment, behaviour, personal, cognitive variables) interact with each other to form a scrupulous pattern of behaviour. Let us take the instance of a student choosing a scrupulous subject for study (behaviour). His/her choice is influenced through his/her family and friends (environment) and also through his/her own personal preference (personal cognitive factor). All these are mutually influencing factors.

Drive Reduction Theory

This theory was presented through Dollard and Miller. When a person is impelled through drives, he learns responses to reduce them. Drive reduction provides rewards and reinforcement. This in turn gives rise to other drives. In this way, the individual learns new responses and new behaviour patterns from our social environment. Consequently his/her personality develops accordingly.

What is the Factors Influencing the Learning Procedure?

They are the abilities of the individual – innate and acquired, his/her stage of maturity, stimulation and the reinforcement he/she gets from the environment.

Application of Control HIV/AIDS

Personality traits are acquired through interacting with the environment. Our lives are not determined merely through our past experiences even though they are significant. Learning processes and their stability help the individual to provide up his/her habits. If a person has acquired a bad habit in sex, it can be changed. The attendance of positive models is also of help in changing our bad habits.

The Humanistic Perspectives on Personality Development

This perspective emphasizes the holistic feature of human experience. This is also recognized as the third force in Psychology, an alternative to Psychodynamic and Behaviour Psychology.

Carl Roger's Self Theory

Carl Rogers proposed the Self-Theory to explain personality development. Everyday and every minute, we are confronted through a number of events in our environment. How do we perceive and interpret these experiences determines our behaviour.

How does personality development occur according to this theory? There are two vital systems underlying Rogers's personality theory – the organism and the self. These two systems are operating in the phenomenological field of an individual. Personality is the product of this interaction.

Phenomenological field means the reality as experienced through an individual or it is the personal reality of the individual. The organism represents the totality of experience (conscious and unconscious). Self is the accepted awareness part of experience.

The acquisition of self concept is a extensive and a continuous procedure. How we perceive our experience – negative or positive – depends upon our self-concept. Self concept is urbanized as a result of a person's interaction with social experience.

For instance, if somebody tells you that you are a very smart and intelligent young boy, you will contain this image in your concept of yourself. You will make every effort to uphold this self-concept.

We regulate our behaviour in order to suit the already shaped self-concepts. What happens if an individual develops a false self image? Inconsistency flanked by real and imagined concepts may result in abnormal behaviour. A healthy personality development occurs if there is a harmonious union of one's self image and his/her real experience in life.

Self Actualization Theory

The mainly well recognized theory on self actualization is Abraham Maslow's theory of hierarchy of needs. He believed that human personality depends upon the fulfilling of inborn potentials. He hypothesized that within every human being, needs are arranged in order of their importance or hierarchy, from the vital to the intricate. He explained five stages of motivational fulfillment of personality development as shown in the diagram below.

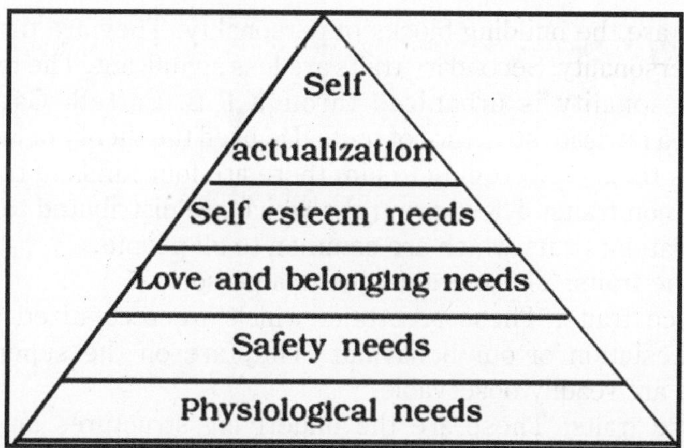

Only after the satisfaction of a need at a lower stage does the after that need become dominant. The satisfaction of primary needs motivates secondary needs. Every individual moves up the steps of the given hierarchy become only after satisfying the vital needs which are essential for one's survival, one can think in relation to the higher needs. A hungry man for instance, cannot think in relation to there forming the society. There are, though, a few exceptions. There are some people who stand for ideals, religious and social values without caring for the satisfaction of their own needs. But such people are very unusual.

Criticism

Still Maslow's need theory has received wide recognition. These theories have contributed much to the individual's freedom, love, personal growth and values. Humanistic theorists have described attention to the constructive side of psychology.

Application in Controlling HIV/AIDS

Humanist psychologists encourage us to see ourselves in terms of positive traits. Troubles occur because of our faulty self image or unsupportive environmental forces. The message from this theory is that through controlling the environment we can control negative exposure to life, especially in unhealthy sexual relations. If we cannot control ourselves we may fall a victim to several-sexual diseases.

Trait Approach

According to the proponents of this approach the vital conceptual unit of personality is the trait. Traits are relatively permanent and constant with a general behaviour pattern that a person exhibits in several situations. G.B. Allport was the first personality theorist to put forward the trait approach to explain personality. He classified traits into cardinal, central and secondary traits. Cardinal traits are very effective and dominant. These are few in number.

Central traits are the building blocks of personality. They are the focus of an individual's personality. Secondary traits are less significant. The mainly recent theory of personality is urbanized through R.B. Cattell. Cattell viewed personality as a intricate structure of traits. He used the theory of factor analysis to develop his theory. According to him there are four kinds of traits.

- Common traits: These are traits which are distributed to the general population; traits which are common to all people.
- Unique traits: These are unique to a person
- Surface traits: These are traits which are recognized through the manifestation of our behaviour. They are on the superficial stage. They are readily observable.
- Source traits: These are the underlying structures or the sources which determine the behaviour of an individual. Source traits are more significant than surface traits. These traits are the vital source of individuality.

Application to Control HIV/AIDS

According to this theory, the source and surface traits are very significant in shaping a personality. The source traits are determined through heredity while surface traits are influenced through environment. Both these traits have a very significant role in controlling HIV/AIDS. The inherited traits should help the individual to control the environment.

DEVELOPMENT OF MORAL VALUES

Every cultural group has its own morals or standards of approved behaviour. Sure acts are measured right because they are conducive to the welfare of its members while sure other acts are wrong as they are dangerous to the welfare of society. What accounts for the emergence of morality in children?

How do they learn the moral values as they grow up? Psychologists have been trying to find answers to these questions for several years. Even today, several psychologists are doing rigorous research in this field.

No one is born with knowledge of what one's cultural group considers right or wrong. As the child grows older, his/her conduct also develops and reflects upon his/ her own moral life as well as that of the group in which the person is a member.

He/she unconsciously begins to imitate the behaviour of others approximately him/her, imitates the customs, manners, traditions and culture of the society, learns that others judge his/her behaviour as right or wrong based on the acceptance of the act through the society. Thus slowly he/she learns the lessons of morality. Learning what the group approves of, is a extensive and a hard procedure.

How do moral values power the personality of an individual?

Moral values directly power the excellence of the person's behaviour and judgment. A person learns the values that satisfy his/her needs and desires and are, at the same time, approved through the group. The moral values acquired through an individual are the reflection of his/her personality. Each individual develops values which are significant for him/her and which guide his/her life.

How do we learn moral values?

There are sure cultural norms of behaviour which are controlled through the society. Learning moral values refers to the manner in which a child internalizes these norms of behaviour. He/she develops moral concepts through learning what is right and what is wrong and also through understanding the use of punishment and other techniques in inculcating moral values in children. The determinants of morality exist on several stages. In every society, there are specific rules to protect morality and to guarantee good behaviour. In addition to this, there are psychological determinants also. Morality has emotional, cognitive and behavioral components. These three components are inter-related. All these factors interact with one another in the development of morality.

THEORIES OF MORAL DEVELOPMENT

In this unit, we shall also talk about the major theories of moral development. Each theory explains only sure characteristics of moral development. Our account highlights the strengths and limitations of each theory.

Socio-biological Theories

The biological perspective of moral development is represented through the field described sociobiology. It assumes that morality is rooted in the genetic heritage of our species. This can be seen from the prosaically behaviour such as helping, co-operating, sharing. This theory cannot explain all characteristics of moral development, but it does point out the adaptive significance of moral behaviour.

Psychoanalytical Theories

Freud's Theory

You have already learnt in relation to the Oedipus intricate arising throughout the phallic stage of psychosexual development. You also know how it can be resolved. Freud assumed that the resolution of Oedipus intricate at the phallic stage marks the beginning of morality. This disagreement arouses fear and anxiety in child. He/she is worried in relation to the loss of parental love. To overcome the anxieties children identify with the same sex parent. The hostility directed towards the same sex parent is redirected towards the self in the form of guilt.

This is the beginning of moral development in children. What are the limitations of this theory? Freud's theory is built approximately guilt as a motivator in moral development. But guilt feeling does not foster conscience development.

Physical punishment and rebukes make child rebellious to power. Instead induction is more effective in conscience formation. Now the question arises, what is induction? It is pointing out to children the consequences of their misbehavior on others. It gives positive instructions to children. This is of immense value in future situations.

Current Psychoanalytic Theories

Recent psychoanalytic ideas place greater importance on the sensitive emotional relationship flanked by parents and children. According to Robert Emde and others (1987) emotional attachment is the vital foundation for acquiring moral standards.

Children who feel secure in their relationship with parents are more likely to learn right behaviour even from early years in life. Parents provide not only prohibitions, but also positive guide rows for behaviour. Both these are incorporated in the super ego of a child. Rewards help the child to internalize the moral standards of parents. For instance, an approving smile from the parent makes him understand the acceptability of an act. Disapproval evokes shame and hurt feelings.

Social Learning Theories

Now let us find out the opinion of the social theorists on moral development. They believe that moral behaviour is acquired presently like any other set of responses through modeling and reinforcement. According to the traditional behaviorist view, children learn new responses through operant conditioning. They learn adult moral standards if their behaviour is reinforced through the mainly significant adult in their life—parents and teachers. The reinforcement can be in the form of praise, approval or rewards. Another way in which children pick up good or bad behaviour is through imitating a role model. The power of the model may last for many weeks after the actual experience. What are the qualities of a model?

Children are more receptive to those personalities who are warm, responsive, powerful and practice what they preach. Punishment is a good motivator for moral actions. But it has limitations. Harsh punishments usually do not promote socially desirable behaviour.

Cognitive Theories

Cognitive theorists assume that individuals develop morally through construction. What is meant through construction in moral development? It means, the procedure through which the child becomes capable of making a correct decision in relation to the rightness or wrongness of an act in a conflicting situation. Through this procedure the child derives new moral insights. They also learn in relation to the principles and procedures for the regulation of human relationships.

Piaget's Theory

One of the earliest theories in this field is propounded through Piaget. Piaget accentuated the importance of rules in moral development. According to him moral development is acquiring the acceptance and respect for rules applied in dissimilar situations.

Piaget's Stages of Moral Development

Piaget recognized two stages of moral development. In the early years of life, children are not concerned in relation to the rules and their implications. But approximately the age of five, they begin to recognize rules and sow respect for whom. This stage is described Heteronomous morality. The word heteronomous means under the power of another. Rules are viewed as external characteristics of society. There is automatic obedience to rules without reasoning or judgment. Why does a child obey rules without questioning? There are several reasons. First of all, these rules are enforced on him through adults who are very powerful people in his life.

The second cause is the intellectual immaturity of the child. He does not know that rules can be customized. The second stage of moral development according to Piaget is Autonomous morality.

At this stage, children use their reasoning power and question the propriety of rules. In 5-12 years age group, a child's concept of justice changes. He/she understands that rules are not rigid, they are flexible and can be revised according to the situation, depending on the will of majority. Peer group power is very prominent at this stage. Today moral development is regarded as a more extended procedure than Piaget believed.

Kohlberg's Extension of Piaget's theory

Kohlberg urbanized a detailed theory on moral development by Piaget's theory as the primary reference. Kohlberg employs a stage and stage approach to describe moral development. According to him there are six moral stages. They are grouped into three major stages consisting of two stages. To understand the stages, an understanding of the three moral stages is essential.

Pre-conventional stage

Moral values reside in external happenings. Moral understanding is based on rewards, punishment and the power of the power figure.

- Stage 1: Punishment Obedience Orientation: Children adopt good behaviour out of fear for punishment. They do not want to get into trouble through disobeying power.
- Stage 2: Instrumental Purpose Orientation: Moral choice is made on the foundation of the satisfaction of personal needs and occasionally the needs of others. They begin to understand that people can have dissimilar perspective in a moral dilemma.

The Conventional Stage

This is the stage of mainly adolescents and adults in our society. You may ask now what is meant through the term conventional. The term conventional means conventionality to the rules, expectations and conventions of the society in which one lives. Moral values at this stage reside in conforming to social rules. Conforming will ensure social order and behaviour according to the expectations of others. It helps the individual to establish positive human relationships.

- Stage 3: Good Boy/Girl Orientation: The desire to get approval and affection from relatives and the peer group motivates good behaviour in children. They want to please and help others. We can also notice the effort on the part of the child to conform to the stereotypical good boy/girl images of the majority.
- Stage 4: The Social Order Maintaining Orientation: In this stage, the individual does his duty to uphold the roles of society. He respects the power. The social order is maintained for one's own sake and for the welfare of the society.

The Post Conventional Stage

This is Kohlberg's highest stage of moral development. At this stage the individual defines morality in terms of abstract principles and values that apply to all situations and societies.

- Stage 5: The Social Contract Orientation: This stage shows a contract flanked by the rights of the individual and the accepted laws of the society. He understands that laws can be changed to protect the rights of the majority. The main emphasis of this stage is community welfare. It focuses on the protection of human rights.
- Stage 6: Universal Ethical Principle Orientation: This stage is conscience oriented. He/she does not strictly adhere to social rules, but is motivated through universal human values. They guide the individual in making moral decisions. The question arising from this

context is: how can we understand the three stages of moral development in the best manner? We can understand the three stages well if we take them as three dissimilar kinds of relationships flanked by the self and society's rules and expectation. At stage I, rules and expectations are something external to the self. At stage II, the person identifies the self with rules and social expectations of others especially those in power. At stage III, the self is differentiated from the rules and expectations of others. He defines his values in terms of self chosen principles.

Criticism

Kohlberg's stages are strongly related to age and form an invariant sequence. But in real life moral development fits a less tightly organized stage conception than the one assumed through Kohlberg. His theory does not take into consideration the full range of moral reasoning in all cultures. This theory underestimates young children's moral understanding.

ENVIRONMENTAL POWERS ON MORAL DEVELOPMENT

What are the factors promoting development of moral values? A few are mentioned below.

- Peer Group Interaction with age mates promotes moral understanding. The child has no inhibitions in engaging in intense, animated discussions in relation to the moral issues with friends.
- Child Rearing Practices: How do child rearing practices adopted through parent's power a child's moral development? Often physical punishments for a wrong act produce negative attitudes in a child. Children are afraid of punishment. To avoid punishment, a child may rationalize his action or may tell lies to escape. The child may even become aggressive with no concern for others. Parents can disapprove of a wrong act and withdraw some privileges given to the child in which case, the child accepts his wrong doing and there is great resistance to temptation. The best method is to have a discussion with the child. Make him understand the consequences of the act as related to himself and to others. Under such circumstances, a mother can power the child more than the father.
- Discipline: Discipline aids the individual in developing moral values. While disciplining children, adults should emphasize the cause why an act is measured right while anther one is wrong. a) Rewards: It has a strong educational value. It shows the child whether or not he/she has behaved in the correct and accepted manner. It motivates him/her to repeat the act. b) Punishment: It powers the development of values only if it is appropriate to the age and action of the child. It

should to administer fairly. Only then will it motivate him/her to conform to social expectations.

- Other Factors: The neighborhood, school mass media, TV, news papers, magazines etc. teach the child culturally approved values of his social group.

Personality Development and Moral Values

The development of personality takes place at home, at school and in society in the attendance of others. The child learns moral values also from these sources. We notice a breakdown in family and social life today. We are faced with several troubles like genetic technology, sexual perversions, HIV/AIDS and so on. These issues threaten the lives of our future generation. Today, in our society, we come crossways several young persons who are proficient in science and technology but ignorant of moral values. In order to equip our youth to fight against all these forces of degradation, we have to inculcate good moral values in them. We should make our young people aware of the importance of good values in life. So, as the child grows older and as his/her personality develops we should acquaint him/her with a moral code accepted through the society.

In today's society learning a moral code is very hard. The reasons are:

- There are a number of dissimilar codes the person encounters within his environment.
- There are inconsistencies flanked by the moral codes and the behaviour of people, and
- Social acceptance of new behaviour patterns brings in relation to the changes in moral codes. These conflicts slow down the development of moral values in an individual. The social group tolerates the violation of the moral codes through young child. But as one grows old one can be punished if the behaviour falls short of social acceptance. If a person realizes that he/ she falls short of the expectations of the society, it may create guilt feelings in the person. The self concept of such a person will be low. Uncertainty, doubt, insecurity will be the personality features if one cannot conform to the values of society and behave accordingly. He/she will be a comfortable and adjusted person.

Moral Development and HIV Control

Cultivating good moral values will prevent the individual from indulging in unhealthy sexual relationships. This will prevent the occurrence of HIV to a sure extent. The society through the institutions of marriage and family regulates the sexual urges of man. Good moral values can help save an individual from the clutches of disease like HIV/AIDS/STDs.

3

Time Management

INTRODUCTION

Time management refers to a range of skills, tools, and techniques used to manage time when accomplishing specific tasks, projects and goals. This set encompasses a wide scope of activities, and these include planning, allocating, setting goals, delegation, analysis of time spent, monitoring, organizing, scheduling, and prioritizing.

Initially time management referred to just business or work activities, but eventually the term broadened to include personal activities as well. A time management system is a designed combination of processes, tools and techniques.

Some authors offered a categorization scheme for the hundreds of time management approaches that they reviewedFirst generation: reminders based on clocks and watches, but with computer implementation possible; can be used to alert a person when a task is to be done.Second generation: planning and preparation based on calendar and appointment books; includes setting goals.

Third generation: planning, prioritizing, controlling activities on a daily basis. This approach implies spending some time in clarifying values and priorities.Fourth generation: being efficient and proactive using any of the above tools; places goals and roles as the controlling element of the system and favors importance over urgency.Some of the recent general arguments related to "time" and "management" point out that the term "time management" is misleading and that the concept should actually imply that it is "the management of our own activities, to make sure that they are accomplished within the available or allocated time, which is an unmanageable continuous resource".

Time management literature paraphrased:

"Get Organized" - paperwork and task triage

"Protect Your Time" - insulate, isolate, delegate

"Set gravitational goals" - that attract actions automatically

"Achieve through Goal management Goal Focus" - motivational emphasis

"Work in Priority Order" - set goals and prioritize

"Use Magical Tools to Get More Out of Your Time" - depends on when written

"Master the Skills of Time Management"

"Go with the Flow" - natural rhythms, Eastern philosophy

"Recover from Bad Time Habits" - recovery from underlying psychological problems, e.g. procrastination

TIME MANAGEMENT AND RELATED CONCEPTS

Time management has been considered as subsets of different concepts such as:

- Project management. Time Management can be considered as a project management subset and is more commonly known as project planning and project scheduling. Time Management has also been identified as one of the core functions identified in project management.
- Attention management: Attention Management relates to the management of cognitive resources, and in particular the time that humans allocate their mind to conduct some activities.
- Personal knowledge management.

PERSONAL TIME MANAGEMENT

Time management strategies are often associated with the recommendation to set goals. These goals are recorded and may be broken down into a project, an action plan, or a simple task list. For individual tasks or for goals, an importance rating may be established, deadlines may be set, and priorities assigned.

This process results in a plan with a task list or a schedule or calendar of activities. Authors may recommend a daily, weekly, monthly or other planning periods, usually fixed, but sometimes variable. Different planning periods may be associated with different scope of planning or review. Authors may or may not emphasize reviews of performance against plan. Routine and recurring tasks may or may not be integrated into the time management plan and, if integrated, the integration can be accomplished in various ways.

Task list

A task list is a list of tasks to be completed, such as chores or steps toward completing a project. It is an inventory tool which serves as an alternative or supplement to memory.Task lists are used in self-management, grocery lists, business management, project management, and software development. It may involve more than one list. When you accomplish one of the items on a task list, you check it off or cross it off. The traditional method is to write these on a piece of paper with a pen or pencil, usually on a note pad or clip-board. Numerous digital equivalents are now available, including PIM applications and most PDAs. There are also several web-based task list applications, many of which are free.

Task list organization

Task lists are often tiered. The simplest tiered system includes a general to-do list to record all the tasks the person needs to accomplish, and a daily to-do list which is created each day by transferring tasks from the general to-do list.

Task lists are often prioritized:

- An early advocate of "ABC" prioritization was Alan Lakein. In his system "A" items were the most important, "B" next most important, "C" least important.
- A particular method of applying the ABC method assigns "A" to tasks to be done within a day, "B" a week, and "C" a month.
- To prioritize a daily task list, one either records the tasks in the order of highest priority, or assigns them a number after they are listed which indicates in which order to execute the tasks. The latter method is generally faster, allowing the tasks to be recorded more quickly.

Alternatives to prioritizing:

A completely different approach which argues against prioritising altogether was put forward by British author Mark Forster in his book "Do It Tomorrow and Other Secrets of Time Management". This is based on the idea of operating "closed" to-do lists, instead of the traditional "open" to-do list. He argues that the traditional never-ending to-do lists virtually guarantees that some of your work will be left undone. This approach advocates getting all your work done, every day, and if you are unable to achieve it helps you diagnose where you are going wrong and what needs to change. Recently, Forster developed the "Autofocus Time Management System", which further systematizes working a to-do list as a series of closed sublists and emphasizes intuitive choices.

Software applications

Modern task list applications may have built-in task hierarchy, may support multiple methods of filtering and ordering the list of tasks, and may allow one to associate arbitrarily long notes for each task. In contrast to the concept of allowing the person to use multiple filtering methods, at least one new software product additionally contains a mode where the software will attempt to dynamically determine the best tasks for any given moment. Many of the software products for time management support multiple users. It allows the person to give tasks to other users and use the software for communication. Task list applications may be thought of as lightweight personal information manager or project management software.

Resistors

- Fear of change: Change can be daunting and one may be afraid to change what's proven to work in the past.
- Uncertainty: Even with the change being inevitable, one may be hesitant as being not sure where to start. Uncertainty about when or how to begin making a change can be significant.
- Time pressure: To save time, one has to invest time, and this time investment may be a cause of concern. Fearing that changing may involve more work at the start—and thus, in the very short term, make things worse—is a common resistor.

Attention Deficit Disorder

Excessive and chronic inability to manage time effectively may be a result of Attention Deficit Disorder.

Diagnostic criteria include: A sense of underachievement, difficulty getting organized, trouble getting started, many projects going simultaneously and trouble with follow-through.

- Prefrontal cortex: The prefrontal cortex is the most evolved part of the brain. It controls the functions of attention span, impulse control, organization, learning from experience and self-monitoring, among others. Some authors argue that changing the way the prefrontal cortex works is possible and offers a solution.

Drivers

- Increased effectiveness: One may feel the need to make more time so as to be more effective in performing the job and carrying out responsibilities.
- Performance improvement: Time management is an issue that often arises during performance appraisals or review meetings.
- Personal development: One may view changing the approach to time management as a personal development issue and reap the benefit of handling time differently at work and at home.
- Increased responsibilities: A change in time-management approach may become necessary as a result of a promotion or additional responsibilities. Since there is more work to do, and still the same amount of time to do it in, the approach must change.
- The end result: working towards the end goal. example completing an assignment.
- Crossing task of the list: The feeling you get when you have completed a task then you cross it off your list, and being able to see what has been completed

Caveats

Dwelling on the lists

- According to Sandberg, task lists "aren't the key to productivity [that] they're cracked up to be". He reports an estimated "30% of listers spend more time managing their lists than [they do] completing what's on them".
- This could be caused by procrastination by prolonging the planning activity. This is akin to analysis paralysis. As with any activity, there's a point of diminishing returns.

Rigid adherence

- Hendrickson asserts that rigid adherence to task lists can create a "tyranny of the to-do list" that forces one to "waste time on unimportant activities".
- Again, the point of diminishing returns applies here too, but toward the size of the task. Some level of detail must be taken for granted for a task system to work. Rather than put "clean the kitchen", "clean the bedroom", and "clean the bathroom", it is more efficient to put "housekeeping" and save time spent writing and reduce the system's administrative load. The risk of consolidating tasks, however, is that "housekeeping" in this example may prove overwhelming or nebulously defined, which will either increase the risk of procrastination, or a mismanaged project.
- Listing routine tasks wastes time. If you are in the habit of brushing your teeth every day, then there is no reason to put it down on the task list. The same goes for getting out of bed, fixing meals, etc. If you need to track routine tasks, then a standard list or chart may be useful, to avoid the procedure of manually listing these items over and over.
- To remain flexible, a task system must allow for disaster. A disaster occurs constantly whether it is personal or business-related. A company must have a cushion of time ready for a disaster. Even if it is a small disaster, if no one made time for this situation, it can blow up bigger, causing the company to bankruptcy just because of poor time management.
- To avoid getting stuck in a wasteful pattern, the task system should also include regular planning and system-evaluation sessions, to weed out inefficiencies and ensure the user is headed in the direction he or she truly desires.
- If some time is not regularly spent on achieving long-range goals, the individual may get stuck in a perpetual holding pattern on short-

term plans, like staying at a particular job much longer than originally planned.

Techniques for setting priorities

ABC analysis

A technique that has been used in business management for a long time is the categorization of large data into groups. These groups are often marked A, B, and C—hence the name. Activities are ranked upon these general criteria:

- A – Tasks that are perceived as being urgent and important.
- B – Tasks that are important but not urgent.
- C – Tasks that are neither urgent nor important.

Each group is then rank-ordered in priority. To further refine priority, some individuals choose to then force-rank all "B" items as either "A" or "C". ABC analysis can incorporate more than three groups. ABC analysis is frequently combined with Pareto analysis.

Pareto analysis

This is the idea that 80% of tasks can be completed in 20% of the disposable time. The remaining 20% of tasks will take up 80% of the time. This principle is used to sort tasks into two parts. According to this form of Pareto analysis it is recommended that tasks that fall into the first category be assigned a higher priority.

The 80-20-rule can also be applied to increase productivity: it is assumed that 80% of the productivity can be achieved by doing 20% of the tasks. If productivity is the aim of time management, then these tasks should be prioritized higher. It depends on the method adopted to complete the task. There is always a simpler and easy way to complete the task. If one uses a complex way, it will be time consuming. So, one should always try to find out the alternate ways to complete each task.

Fit

Essentially, fit is the congruence of the requirements of a task with the available resources at the time. Often people are constrained by externally controlled schedules, locations, etc., and "fit" allows us to maximize our productivity given those constraints. For example, if one encounters a gap of 15 minutes in their schedule, it is typically more efficient to complete a task that would require 15 minutes, than to complete a task that can be done in 5 minutes, or to start a task that would take 4 weeks. This concept also applies to time of the day: free time at 7am is probably less usefully applied to the goal of learning the drums, and more productively a time to read a book. Lastly, fit can be applied to location: free time at home would be used differently from free time at work, in town, etc.

POSEC method

POSEC is an acronym for Prioritize by Organizing, Streamlining, Economizing and Contributing.

The method dictates a template which emphasizes an average individual's immediate sense of emotional and monetary security. It suggests that by attending to one's personal responsibilities first, an individual is better positioned to shoulder collective responsibilities.

Inherent in the acronym is a hierarchy of self-realization which mirrors Abraham Maslow's "Hierarchy of needs".

1. Prioritize - Your time and define your life by goals.
2. Organizing - Things you have to accomplish regularly to be successful.
3. Streamlining - Things you may not like to do, but must do.
4. Economizing - Things you should do or may even like to do, but they're not pressingly urgent.
5. Contributing - By paying attention to the few remaining things that make a difference.

The Eisenhower Method

All tasks are evaluated using the criteria important/unimportant and urgent/not urgent and put in according quadrants. Tasks in unimportant/not urgent are dropped, tasks in important/urgent are done immediately and personally, tasks in unimportant/urgent are delegated and tasks in important/not urgent get an end date and are done personally. This method is said to have been used by US President Dwight D. Eisenhower, and is outlined in a quote attributed to him: What is important is seldom urgent and what is urgent is seldom important.

PROJECT MANAGEMENT

Project Management is a well-established approach to managing and controlling the introduction of new initiatives or organizational changes. Projects are finite in length, usually one-time pieces of work involving a number of activities that must be completed within a given time frame, and often on a fixed budget. Common examples of projects are construction of a building, introduction of a new product, installation of a new piece of machinery in a manufacturing plant, creation of a new software tool, or the design and launch of a new advertising campaign.

While the very simplest projects can be managed easily by applying common sense and just getting on with things, projects that are more complex need a great deal of planning, and benefit from a formal, disciplined management approach. From making sure that activities will actually meet the specified need, to devising a workable schedule, developing systems for reporting progress, and managing requests for changes – all of these issues require thoughtful consideration. Managing projects well requires a great deal of time, skill, and finesse. There are many sides to project management and this is what makes it so interesting and demanding. Project managers are expected to take an uncertain event and make a certain promise to deliver. They are also expected to do this within a specified time and within a limited budget.

ATTENTION MANAGEMENT

Attention management refers to models and tools for supporting the management of attention at the individual or at the collective level and at the short term of at a longer terms.

APPROACHES FOR ATTENTION SUPPORT

Different researches have been conducted in using ICT for supporting attention, and in particular models have been elaborated for supporting attention. Supporting the management of attention objective is to bring a certain number of solution to:

- people perception cognitive limitations, such as the limited capacity of the human short term memory or the theoretical cognitive limit to the number of people with whom one can maintain stable social relationships.
- information overload
- social interaction overload
- interruption
- multitasking

Tools can been designed for supporting attention

- at the organizational level, by supporting organisation processes
- at the collective level
- at the individual level, for instance using attentive user interfaces.
- at the individual level by helping people to assess & analyse their attention related practices.

These tools are often adaptive hypermedia, and often rely on the profiling the user in order determine how to better support people attention.

PROJECTS

A certain number of projects have been conducted to investigate how to use ICT to support attention such as:

- AtGentive Attentive Agents for Collaborative Learners.
- SAKE - Semantic-enabled Agile Knowledge-based eGovernment
- SUITOR

PERSONAL KNOWLEDGE MANAGEMENT

Personal knowledge management refers to a collection of processes that an individual carries out to gather, classify, store, search, retrieve, and share knowledge in his/her daily activities and how these processes support work activities. It is a response to the idea that knowledge workers increasingly need to be responsible for their own growth and learning and represents a bottom-up approach to knowledge management, as opposed to more traditional, top-down KM.

Although as early as 1998 Davenport wrote on the importance to worker productivity of understanding individual knowledge processes, the term personal knowledge management appears to be relatively new. Its origin can be traced in a working paper by Frand and Hixon.

PKM integrates personal information management, focused on individual skills, with knowledge management in addition to input from a variety of disciplines such as cognitive psychology, management, and philosophy. From an organizational perspective, understanding of the field has developed in light of expanding knowledge about human cognitive capabilities and the permeability of organizational boundaries. From a metacognitive perspective, it compares various modalities within human cognition as to their competence and efficacy. It is an underresearched area.

PKM MODELS

Dorsey identified information retrieval, assessment and evaluation, organization, analysis, presentation, security, and collaboration as essential to PKM. Wright's model involves four interrelated domains: analytical, information, social, and learning. The analytical domain involves competencies such as interpretation, envisioning, application, creation, and contextualization. The information dimension comprises the sourcing, assessment, organization, aggregation, and communication of information. The social dimension involves finding and collaborating with people, development of both close networks and extended networks, and dialogue. The learning dimension entails expanding pattern recognition and sensemaking capabilities, reflection, development of new knowledge, improvement of skills, and extension to others. This model stresses the importance of both bonding and bridging networks.

In Nonaka and Takeuchi's SECI model of knowledge creation, knowledge can be tacit or explicit, with the interaction of the two resulting in new knowledge. Smedley has developed a PKM model based on Nonaka and colleagues' model in which an expert provides direction and a community of practice provides support for personal knowledge creation. Trust is central to knowledge sharing in this model.

Personal knowledge management can also be viewed along two main dimensions, personal knowledge and personal management. Zhang has developed a model of PKM in relation to organizational knowledge management that considers two axes of knowledge properties and management perspectives, either organizational or personal. These aspects of organizational and personal knowledge are interconnected through the OAPI process, whereby organizational knowledge is personalized and individualized and personal knowledge is aggregated and operationalized as organizational knowledge.

CRITICISMS OF PKM

Not everyone agrees that the focus on the individual is a good thing, or that PKM is anything more than a new wrapper around personal information management. Dave Snowden has asserted that most individuals cannot manage their knowledge in the traditional sense of "managing" and has advocated thinking in terms of sensemaking rather than PKM. Most notably, some argue that knowledge is never an individual product - that it emerges through connections, dialog and social interaction. However, in Wright's model, PKM involves the application to problem solving of analytical, information, social, and learning dimensions, which are interrelated, and so is inherently social.

An aim of PKM is "helping individuals to be more effective in personal, organisational and social environments", often through the use of technology such as networking software. It has been argued, however, that equation of PKM with technology has limited the value and utility of the concept.

PKM SKILLS

Skills associated with personal knowledge management.

- Reflection. Continuous improvement on how the individual operates.
- Manage learning. Manage how and when the individual learns.
- Information literacy. Understanding what information is important and how to find unknown information.
- Organizational skills. Personal librarianship. Personal categorization and taxonomies.
- Networking with others. Knowing what your network of people knows. Knowing who might have additional knowledge and resources to help you
- Researching, canvassing, paying attention, interviewing and observational 'cultural anthropology' skills
- Communication skills. Perception, intuition, expression, visualization, and interpretation.
- Creative skills. Imagination, pattern recognition, appreciation, innovation, inference. Understanding of complex adaptive systems.
- Collaboration skills. Coordination, synchronization, experimentation, cooperation, and design.

PKM TOOLS

Some organizations are now introducing PKM 'systems' with some or all of four components:

- Just-in-time Canvassing - templates and e-mail canvassing lists that enable people to identify and connect with the appropriate experts and expertise quickly and effectively

- Knowledge harvesting - software tools that automatically collect appropriate knowledge residing on subject matter experts' hard drives
- Content management tools - taxonomy processes and desktop search tools that enable employees to subscribe to, find, organize, and publish information that resides on their desktops
- Personal Productivity Improvement - knowledge fairs and one-on-one training sessions to help each employee make more effective personal use of the knowledge, learning, and technology resources available in the context of their work

PKM has also been linked to these tools:

- social bookmarking and enterprise social bookmarking
- knowledge logs
- e-mail, calendars, task managers
- Online Web Assistants
- Wikis
  - Desktop wikis
  - Personal wikis
  - Semantic Wikis

Other useful tools include Open Space Technology, cultural anthropology, stories and narrative, mindmaps, concept maps and eco-language, and single frames and similar visualization techniques. Individuals use these tools to capture and ideas, expertise, experience, opinions, or thoughts, and this 'voicing' will encourage cognitive diversity and promote free exchanges away from a centralized policed knowledge repository. The goal is to facilitate knowledge sharing and personal content management.

BEATING PROCRASTINATION

If you've found yourself putting off important tasks over and over again, you're not alone. In fact, many people procrastinate to some degree – but some are so chronically affected by procrastination that it stops them fulfilling their potential and disrupts their careers. The key to controlling this destructive habit is to recognize when you start procrastinating, understand why it happens and take active steps to manage your time and outcomes better.

WHAT IS PROCRASTINATION?

In a nutshell, you procrastinate when you put off things that you should be focusing on right now, usually in favor of doing something that is more enjoyable or that you're more comfortable doing. According to psychologist Professor Clarry Lay, a prominent writer on procrastination, procrastination occurs when there's "a temporal gap between intended behavior and enacted behavior." That is, procrastination is occurring when there's a significant time period between when people intend to do a job, and when they actually do it.

HOW TO OVERCOME PROCRASTINATION

Follow these steps to deal with and control procrastination:

Step 1: Recognize That You're Procrastinating

If you're honest with yourself, you probably know when you're procrastinating. But to be sure, take our Are You a Procrastinator? self test.

Here are some useful indicators that will help you know when you're procrastinating:

- Filling your day with low priority tasks from your To Do List.
- Reading e-mails several times without starting work on them or deciding what you're going to do with them.
- Sitting down to start a high-priority task, and almost immediately going off to make a cup of coffee.
- Leaving an item on your To Do list for a long time, even though you know it's important.
- Regularly saying "Yes" to unimportant tasks that others ask you to do, and filling your time with these instead of getting on with the important tasks already on your list.
- Waiting for the "right mood" or the "right time" to tackle the important task at hand.

Step 2: Work Out WHY You're Procrastinating

Why you procrastinate can depend on both you and the task. But it's important to understand which of the two is relevant in a given situation, so that you can select the best approach for overcoming your reluctance to get going. One reason for procrastination is that people find a particular job unpleasant, and try to avoid it because of that. Most jobs have unpleasant or boring aspects to them, and often the best way of dealing with these is to get them over and done with quickly, so that you can focus on the more enjoyable aspects of the job.

Another cause is that people are disorganized. Organized people manage to fend of the temptation to procrastinate, because they will have things like prioritized to-do lists and schedules which emphasize how important the piece work is, and identify precisely when it's due. They'll also have planned how long a task will take to do, and will have worked back from that point to identify when they need to get started in order to avoid it being late. Organized people are also better placed to avoid procrastination, because they know how to break the work down into manageable "next steps".

Even if you're organized, you can feel overwhelmed by the task. You may doubt that you have the skills or resources you think you need, so you seek comfort in doing tasks you know you're capable of completing. Unfortunately, the big task isn't going to go away – truly important tasks rarely do. You may also fear success as much as failure. For example, you may think that success will lead to you being swamped with more requests to do this type of task, or that you'll be pushed to take on things that you feel are beyond you.

Surprisingly, perfectionists are often procrastinators, as they can tend to think "I don't have the right skills or resources to do this perfectly now, so I won't do it at all." One final major cause of procrastination is having underdeveloped decision-making skills. If you simply can't decide what to do, you're likely to put off taking action in case you do the wrong thing.

Step 3: Adopt Anti-Procrastination Strategies

Procrastination is a habit – a deeply ingrained pattern of behavior. That means that you won't just break it overnight. Habits only stop being habits when you have persistently stopped practising them, so use as many approaches as possible to maximize your chances of beating procrastination. Some tips will work better for some people than for others, and for some tasks than others. And, sometimes, you may simply need to try a fresh approach to beat the "procrastination peril"!

These general tips will help motivate you to get moving:

- Make up your own rewards. For example, promise yourself a piece of tasty flapjack at lunchtime if you've completed a certain task. And make sure you notice how good it feels to finish things!
- Ask someone else to check up on you. Peer pressure works! This is the principle behind slimming and other self-help groups, and it is widely recognized as a highly effective approach.
- Identify the unpleasant consequences of NOT doing the task.
- Work out the cost of your time to your employer. As your employers are paying you to do the things that they think are important, you're not delivering value for money if you're not doing those things. Shame yourself into getting going!
- Aim to "eat an elephant beetle" first thing, every day!

If you're pocrastinating because you're disorganized, here's how to get organized!

- Keep a To-Do list so that you can't "conveniently" forget about unpleasant or overwhelming tasks.
- Use an Urgent/Important Matrix to help prioritize your to-do list so that you can't try to kid yourself that it would be acceptable to put off doing something on the grounds that it's unimportant, or that you have many urgent things which ought to be done first when, in reality, you're procrastinating.

- Become a master of scheduling and project planning, so that you know when to start those all-important projects.
- Set yourself time-bound goals: that way, you'll have no time for procrastination!
- Focus on one task at a time.

If you're putting off starting a project because you find it overwhelming, you need to take a different approach. Here are some tips:

- Break the project into a set of smaller, more manageable tasks. You may find it helpful to create an action plan.
- Start with some quick, small tasks if you can, even if these aren't the logical first actions. You'll feel that you're achieving things, and so perhaps the whole project won't be so overwhelming after all.

If you're procrastinating because you find the task unpleasant:

- Many procrastinators overestimate the unpleasantness of a task. So give it a try! You may find that it's not as bad as you thought!
- Hold the unpleasant consequences of not doing the work at the front of your mind.
- Reward yourself for doing the task.

Finally, if you're procrastinating because you can't decide what action to take, and are putting off making a decision because you're nervous about making the wrong choice, see our decision-making section. This teaches a range of powerful and effective decision-making techniques. Remember: the longer you can spend without procrastinating, the greater your chances of breaking this destructive habit for good!

KEY POINTS

To have a good chance of conquering procrastination, you need to spot straight away that you're doing it. Then, you need to identify why you're procrastinating and taken appropriate steps to overcome the block. Part of the solution is to develop good time management, organizational and personal effectiveness habits, such as those described in Make Time for Success! This helps you establish the right priorities, and manage your time in such a way that you make the most of the opportunities open to you.

ACTIVITY LOGS

How long do you spend each day on unimportant things; Things that don't really contribute to your success at work? Do you KNOW how much time you've spent reading junk mail, talking to colleagues, making coffee and eating lunch? And how often have you thought, "I could achieve so much more if I just had another half hour each day." And are you aware of when in the day you check your e-mail, write important articles or do your long-term planning? Most people find they function at different levels of effectiveness at different times of day as their energy levels fluctuate. Your effectiveness may vary depending on the amount of sugar in your blood, the length of time since you last took a break, routine distractions, stress, discomfort, or a range of other factors.

Activity logs help you to analyze how you actually spend your time. The first time you use an activity log you may be shocked to see the amount of time that you waste! Memory is a very poor guide when it comes to this, as it can be too easy to forget time spent on non-core tasks.

Keeping an Activity Log for several days helps you to understand how you spend your time, and when you perform at your best. Without modifying your behavior any further than you have to, note down the things you do as you do them on this template. Every time you change activities, whether opening mail, working, making coffee, gossiping with colleagues or whatever, note down the time of the change.

As well as recording activities, note how you feel, whether alert, flat, tired, energetic, etc. Do this periodically throughout the day. You may decide to integrate your activity log with a stress diary.Once you have logged your time for a few days, analyze your daily activity log. You may be alarmed to see the amount of time you spend doing low value jobs!

You may also see that you are energetic in some parts of the day, and flat in other parts. A lot of this can depend on the rest breaks you take, the times and amounts you eat, and quality of your nutrition. The activity log gives you some basis for experimenting with these variables.

Your analysis should help you to free up extra time in your day by applying one of the following actions to most activities:

- Eliminate jobs that your employer shouldn't be paying you to do. These may include tasks that someone else in the organization should be doing, possibly at a lower pay rate, or personal activities such as sending non-work e-mails.
- Schedule your most challenging tasks for the times of day when your energy is highest. That way your work will be better and it should take you less time.
- Try to minimize the number of times a day you switch between types of task. For example, read and reply to e-mails in blocks once in the morning and once in the afternoon only.
- Reduce the amount of time spent on legitimate personal activities such as making coffee.

Activity logs are useful tools for auditing the way that you use your time. They can also help you to track changes in your energy, alertness and effectiveness throughout the day. By analyzing your activity log you will be able to identify and eliminate time-wasting or low-yield jobs. You will also know the times of day at which you are most effective, so that you can carry out your most important tasks during these times.

TO-DO LISTS

Do you frequently feel overwhelmed by the amount of work you have to do? Do you face a constant barrage of looming deadlines? Or do you sometimes just forget to do something important, so that people have to chase you to get work done?

All of these are symptoms of not keeping a proper "To-Do List". To-Do Lists are prioritized lists of all the tasks that you need to carry out. They list everything that you have to do, with the most important tasks at the top of the list, and the least important tasks at the bottom. While this sounds a simple thing to do, it's when people start to use To-Do Lists properly that they often make their first personal productivity/ time management breakthrough, and start to make a real success of their careers.

By keeping a To-Do List, you make sure that you capture all of the tasks you have to complete in one place. This is essential if you're not going to forget things. And by prioritizing work, you plan the order in which you'll do things, so you can tell what needs your immediate attention, and what you can quietly forget about until much, much later. This is essential if you're going to beat work overload. Without To-Do Lists, you'll seem dizzy, unfocused and unreliable to the people around you. With To-Do Lists, you'll be much better organized, and will seem much more reliable. This is very important!

PREPARING A TO-DO LIST

Start by writing down all of the tasks that you need to complete, and if they are large, break them down into their component elements. If these still seem large, break them down again. Do this until you have listed everything that you have to do, and until tasks are will take no more than 1-2 hours to complete. This may be a huge and intimidating list, but our next step makes it manageable!

Next, run through these jobs allocating priorities from A to F. If too many tasks have a high priority, run through the list again and demote the less important ones. Once you have done this, rewrite the list in priority order. You will then have a precise plan that you can use to eliminate the problems you face. You will be able to tackle these in order of importance or urgency. This allows you to separate important jobs from the many time-consuming trivial ones.

USING YOUR TO-DO LISTS

Different people use To-Do Lists in different ways in different situations: if you are in a sales-type role, a good way of motivating yourself is to keep your list relatively short and aim to complete it every day. In an operational role, or if tasks are large or dependent on too many other people, then it may be better to keep one list and 'chip away' at it.

It may be that you carry unimportant jobs from one To-Do List to the next. You may not be able to complete some very low priority jobs for several months. Only worry about this if you need to – if you are running up against a deadline for them, raise their priority. If you have not used To-Do Lists before, try them now: They are one of the keys to being really productive and efficient.

KEY POINTS

Prioritized To-Do Lists are fundamentally important to efficient work. If you use To-Do Lists, you will ensure that:

- You remember to carry out all necessary tasks
- You tackle the most important jobs first, and do not waste time on trivial tasks.
- You do not get stressed by a large number of unimportant jobs.

To draw up a Prioritized To-Do List, download our template and use it to list all the tasks you must carry out. Mark the importance of the task next to it, with a priority from A to F. Redraft the list into this order of importance. Now carry out the jobs at the top of the list first. These are the most important, most beneficial tasks to complete.

ACTION PROGRAMS

To-Do Lists are great for managing a small number of tasks. The problem is that, for most of us, our To-Do List is not really a planned, focused action list. Rather, it is a sort of a catch-all for a lot of things that are unresolved and not yet translated into outcomes. Specific entries, such as "Call Tina," exist along with vaguer aspirations, such as "Get started on house painting project." Often, the real actionable details of what the list-maker has "to do" are actually missing. What this means is that you tend to do the specific tasks, and fail to make progress with the big, important projects. And even if you do get beyond the quick actions, having a complete project as a "to do" can lead you to focus all of your attention on it. This makes multi-tasking difficult. This can be a serious problem in a job where you need to make progress on many different projects at the same time - and this is exactly the situation most senior managers find themselves in.

This is where Action Programs are useful. Action Programs are "industrial strength" versions of To-Do Lists. Because they incorporate short-, medium- and long-term goals, they allow you to plan your time, without forgotten commitments coming in to blow your schedule apart. Because priorities are properly thought through, you'll be focusing on the things that matter, and not frittering your time away on low value activities. And because they support delegation, they help you get into the habit of delegating jobs where you can. All of this lets you save time - and get away on time - whilst also significantly increasing your effectiveness and productivity.

HOW TO USE THE TOOL

Follow this four-step procedure to create your Action Program:

Step 1. Collection

First, make a list inventory of all the things in your world that require resolution. Try to collect and write down everything - urgent or not, big or small, personal or professional - that you feel is incomplete and needs action from you to get completed. To an extent, this collection is taking place automatically. E-mail requests are getting stored in your email account, memos demanding attention are being delivered to your in-tray, mail is reaching your mailbox and messages asking for action are accumulating on your voice mail. But there is other stuff - stuff that is idling in your head, projects you want to run, things you intend to deal with lying at the bottom of the drawer, ideas written down on stray bits of paper - that need to be gathered and put in place too. Bring all of these actions and projects together and inventory them in one place. And - this is really important - make sure that your personal goals are brought onto this list.

Step 2. Pruning

Now, process the list you made in step 1, by looking carefully at each item.

Decide whether you should, actually, take action on it. A lot of what comes our way has no real relevance to us, or is really not important in the scale of things. If that is the case, then delete these things from your inventory.

Step 3. Organizing and Prioritizing

This comes in three parts.

First of all, review your inventory of items. For any which are separate, individual actions that make up part of a larger project, group these individual actions together into their projects.

For example, at home, you may want to improve your bathroom, and repaint your living room: these can go into a "Home Renovation" project. At work, you may be providing input into the requirements for a new computer system, and may be expected to test and then train your team on this system at a point in the future: all of these go into a "computer system" project. What you'll find is that once you start, items will almost seem to "organize themselves" into coherent projects.

You also need to make sure that your personal goals are included as individual projects. Second, review these projects, and allocate a priority to them depending on their importance. Clearly, your personal goals are exceptionally important projects! Third, insert your projects into a formatted Action Program.

The Action Program is split up into three parts:

1. A "Next Action List," which shows the small next actions that you will take to move your projects forward.
2. A "Delegated Actions List," which shows projects and actions have delegated to other people.
3. A "Project Catalog" that shows all of the projects you are engaged in and the small individual tasks that you have identified so far that contribute to them.

The great news is that, by this stage, you've already created the largest part of this: the Project Catalog! This is the list of prioritized projects and activities that you've just completed. Typically, the Project Catalog is at the back of the Action Program, as it's often only referred to during a weekly review process.

Next, create the Delegated Actions List by working through your Project Catalog, and identifying tasks that you've delegated. Record these under the name of the person who you've delegated the activity to, along with the checkpoints you've agreed. Typically, the Delegated Actions List sits in front of the Project Catalog in your Action Program document, as it's referred to quite often. Finally, create your Next Action List by working through the projects to which you've given the highest priority - the projects that you want and need to move forward on straight away - and extracting the small, logical next actions for these projects.

The Next Action List goes on the front page of your Action Program, as you'll refer to it many times a day. Now review the Next Action List. If it is too cluttered, move some of the less urgent/important jobs back into the project catalog. If it is thin and under-challenging, pull up some more Next Actions from the Project Catalog. Also, it makes sense to prioritize the items in the Next Action List so you know what to focus on.

Step 4. "Working" Your Action Program

An Action Program is typically fairly long. But you don't have to run through the entire Program every day! Usually, you'll only be dealing with the top page or pages. Some activities may be day-specific or time-specific. Depending of the way you work, these can be either maintained as the top page of your Action Program or marked on your calendar.

In effect, these pages are just a new form of your old To-Do List. It is just that only specific short actions are outlined here, while the major projects to which the actions belong are stored in your Project Catalog. What you must do, however, is review your Action Program periodically, for example, every week. Delete or archive items you've completed, move items from the Project Catalog to the front pages as you make progress on your project, and add any new actions that have come your way.

PRIORITIZATION

Prioritization is the essential skill you need to make the very best use of your own efforts, and those of your team. It is particularly important when time is limited and demands are seemingly unlimited. It helps you to spend your time wisely, freeing you and your team up from less important tasks that can be attended to later – or quietly dropped.

With good prioritization you can bring order to chaos, massively reduce stress, and move forwards successfully. Without it, you'll flounder around, drowning in competing demands.

SIMPLE PRIORITIZATION

At a simple level, you can prioritize based on time constraints, on the potential profitability or benefit of the task you're facing, or on the pressure you're under to complete a job:

- Prioritization based on project value or profitability is probably the most commonly-used and rational basis for prioritization. Whether this is based on a subjective guess at value or a sophisticated financial evaluation, it often gives the most efficient results.
- Time constraints are important where other people are depending on you to complete a task, and particularly where this task is on the critical path of an important project. Here, a small amount of your own effort can go a very long way.
- And it's a brave person who resists his or her boss's pressure to complete a task, when that pressure is reasonable and legitimate.

PRIORITIZATION TOOLS

While these simple approaches to prioritization suit many situations, there are plenty of special cases where you'll need other prioritization and time management tools if you're going to be truly effective. We look at some of these below: While these simple approaches to prioritization suit many situations, there are plenty of special cases where you'll need other tools if you're going to be truly effective. We look at some of these below:

- Paired Comparison Analysis: Paired Comparison Analysis is most useful where decision criteria are vague, subjective or inconsistent. It helps you prioritize options by asking you to compare each item on a list with all other items on the list individually. By deciding in each case which of the two is most important, you can consolidate results to get a prioritized list. Click here to find out more about Paired Comparison Analysis.
- Grid Analysis: Grid Analysis helps you prioritize a list of tasks where you need to take many different factors into consideration. Click here to learn how to use it.

- The Action Priority Matrix: This quick and simple diagramming technique asks you to plot the value of the task against the effort it will consume. By doing this you can quickly spot the "quick wins" which will give you the greatest rewards in the shortest possible time, and avoid the "hard slogs" which soak up time for little eventual reward. This is an ingenious approach for making highly efficient prioritization decisions. Click here to find out more.
- The Urgent/Important Matrix:Similar to the Action Priority Matrix, this technique asks you to think about whether tasks are urgent or important. Frequently, seemingly urgent tasks actually aren't that important. And often, really important activities just aren't that urgent. This approach helps you cut through this. Click here to find out more.
- The Ansoff & Boston Matrices: These give you quick "rules of thumb" for prioritizing the opportunities open to you. The Ansoff Matrix helps you evaluate and prioritize opportunities by risk. The Boston Matrix does a similar job, helping you prioritize opportunities based on the attractiveness of a market and your ability to take advantage of it. For more information on the Ansoff Matrix, click here: And for the Boston Matrix, see here.
- Pareto Analysis: Where you're facing a flurry of problems needing to be solved, Pareto Analysis helps you identify the most important changes to make. It firstly asks you to group together the different types of problem you face, and then asks you to count the number of cases of each type of problem. By prioritizing the most common type of problem, you can focus your efforts on resolving it. This clears time to focus on the next set of problems, and so on.
- Nominal Group Technique: Nominal Group Technique is a useful technique for prioritizing issues and projects within a group, giving everyone fair input into the prioritization process. This is particularly useful where consensus is important, and where a robust group decision needs to be made. Using this tool, each group participant "nominates" his or her priority issues, and then ranks them on a scale, of say 1 to 10. The score for each issue is then added up, with issues then prioritized based on scores. The obvious fairness of this approach makes it particularly useful where prioritization is based on subjective criteria, and where people's "buy in" to the prioritization decision is needed.

EFFECTIVE SCHEDULING

So far in this section of Mind Tools, we have looked at your priorities and your goals – these define what you aspire to do with your time. Scheduling is where these aspirations meet reality.

Scheduling is the process by which you look at the time available to you, and plan how you will use it to achieve the goals you have identified. By using a schedule properly, you can:

- Understand what you can realistically achieve with your time;
- Plan to make the best use of the time available;
- Leave enough time for things you absolutely must do;
- Preserve contingency time to handle 'the unexpected'; and
- Minimize stress by avoiding over-commitment to others.

A well thought-through schedule allows you to manage your commitments, while still leaving you time to do the things that are important to you. It is therefore your most important weapon for beating work overload.

HOW TO USE THE TOOL

Scheduling is best done on a regular basis, for example at the start of every week. Go through the following steps in preparing your schedule:

1. Start by identifying the time you want to make available for your work. This will depend on the design of your job and on your personal goals in life.
2. Next, block in the actions you absolutely must take to do a good job. These will often be the things you are assessed against. For example, if you manage people, then you must make time available for coaching, supervision, and dealing with issues that arise. Similarly, you must allow time to communicate with your boss and key people around you.
3. Review your To Do List, and schedule in the high-priority, urgent activities, as well as the essential maintenance tasks that cannot be delegated and cannot be avoided.
4. Next, block in appropriate contingency time. You will learn how much of this you need by experience. Normally, the more unpredictable your job, the more contingency time you need. The reality of many people's work is of constant interruption: Studies show some managers getting an average of as little as six minutes uninterrupted work done at a time. Obviously, you cannot tell when interruptions will occur. However, by leaving space in your schedule, you give yourself the flexibility to rearrange your schedule to react effectively to urgent issues.
5. What you now have left is your "discretionary time": the time available to deliver your priorities and achieve your goals. Review your Prioritized To Do List and personal goals, evaluate the time needed to achieve these actions, and schedule them in.

By the time you reach step 5, you may find that you have little or no discretionary time available. If this is the case, then revisit the assumptions you used in the first four steps. Question whether things are absolutely necessary, whether they can be delegated, or whether they can be done in an abbreviated way.

Remember that one of the most important ways people learn to achieve success is by maximizing the 'leverage' they can achieve with their time. They increase the amount of work they can manage by delegating work to other people, spending money outsourcing key tasks, or using technology to automate as much of their work as possible. This frees them up to achieve their goals.

Also, use this as an opportunity to review your To Do List and Personal Goals. Have you set goals that just aren't achievable with the time you have available? Are you taking on too many additional duties? Or are you treating things as being more important than they really are? If your discretionary time is still limited, then you may need to renegotiate your workload. With a well-thought through schedule as evidence, you may find this surprisingly easy.

KEY POINTS

Scheduling is the process by which you plan your use of time. By scheduling effectively, you can reduce stress and maximize your effectiveness. This makes it one of the most important time management skills you can use. Before you can schedule efficiently, you need an effective scheduling system. This can be a diary, calendar, paper-based organizer, PDA or a software package like MS Outlook. The best solution depends entirely on your circumstances.

Scheduling is then a five-step process:

1. Identify the time you have available.
2. Block in the essential tasks you must carry out to succeed in your job.
3. Schedule in high priority urgent tasks and vital "house-keeping" activities.
4. Block in appropriate contingency time to handle unpredictable interruptions.
5. In the time that remains, schedule the activities that address your priorities and personal goals.

If you have little or no discretionary time left by the time you reach step five, then revisit the assumptions you have made in steps one to four.

GOAL SETTING

The ability to set goals is at the very core of personal productivity, career success, and effective leadership; and the resources on this page help you learn how to do this. Explore the articles below, and start setting – and achieving – your goals today!

PERSONAL GOAL SETTING

Why is personal goal setting so important in personal time management? From the time management perspective, your life is a sequence of big and small choices and decisions.

It is those choices that you really manage, not the flow of time. Personal goal setting is the wisdom that comes out of a lot of practical experience and psychology research to help you direct your conscious and subconscious decisions towards success, building up your motivation to achieve your personal or business goals.

LOCKE'S GOAL SETTING THEORY

Edwin Locke's theory of goal setting has many practical applications, in and outside of the business environment. A professor at the University of Maryland, Locke's theory defines characteristics that encourage success. While the theory rings of psychology, its applications in the business world have been profound and enduring.

The Influence of Ryan

Locke's goal-setting theory was created on the premise, originally set forth by professor Thomas A. Ryan, that "conscious goals affect action." As quoted in Locke's "Building a Practically Useful Theory of Goal Setting and Task Motivation," Ryan argued that human behavior is affected by conscious purposes, plans, intentions, tasks and the like.

Basic Definition

Locke's theory operates on the premise that individuals create goals by making careful decisions to do so and are compelled toward those goals by virtue of the goal having been set. Basically, Locke's theory states that if an individual sets goals, he will be motivated to achieve those goals by virtue of having set them.

Several elements must exist in order for the goal-setting effect to take place. Goals must be clear, challenging and attainable, and there must be some method of receiving feedback. Locke finds that the goal itself is not the motivator, but rather the perceived difference between what was actually attained and what had been planned for.

Goal Difficulty and Performance

Locke, and professors Steve Motowidlo and Phil Bobko found that "higher expectancies lead to higher levels of performance," which is in accordance with Vroom's valence-instrumentality-expectancy theory. Somewhat contradictorily, they also showed that when expectations are low but goal level is high, performance would be high also.

Goal Mechanisms

Goals serve four primary functions:

1. By specifying a goal, one must direct focus toward that goal and away from activities unrelated to that goal.
2. The setting of a goal is a behavior-stimulating act. According to Locke, "high goals lead to greater effort than low goals."
3. Goals have a positive effect on persistence. However, there is an inverse relationship between time and intensity.
4. Goals subconsciously direct the person toward discovering better ways of doing things, be they calculations or physical acts.

Goal Moderators

Locke's theory states that, in order for a goal to be successful, the person must be committed to it wholly and possess self-efficacy. This self-efficacy must be boosted initially by the fact that the person was assigned the task and thus believed to be capable of its completion. He also found that "for goals to be effective, people need summary feedback that reveals progress in relation to their goals. If they do not know how they are doing, it is difficult or impossible for them to adjust the level or direction of their effort or to adjust their performance strategies to match what the goal requires." Task complexity also moderates the effects of goals because more complex goals require the review of more complex strategies than lower difficulty goals. Lastly, more complex goals require proximal goals rather than a singular distal goal. Basically, the complex goals should be broken down into several smaller goals. The setting of proximal goals also promotes progress feedback.

Limitations

As noted by Locke, his goal-setting theory has several limitations:

1. Goal conflict. Sometimes an individual has several goals, some of which may be in conflict. When this occurs, performance will suffer.
2. Goals and risk. More difficult goals/deadlines can spur riskier behaviors and strategies.
3. Personality. Goal success is largely effected by self-efficacy. Also, personality plays a large role in goal determination and approach.
4. Goals and subconscious motivation. Subconscious motivators affect people regularly, but how these subconscious motivators affect goal performance has not been studied.

GOLDEN RULES OF GOAL SETTING

If you want to succeed, you need to set goals. Without goals you lack focus and direction. Goal setting not only allows you to take control of your life's direction; it also provides you a benchmark for determining whether you are actually succeeding. Think about it: Having a million dollars in the bank is only proof of success if one of your goals is to amass riches. If your goal is to practice acts of charity, then keeping the money for yourself is suddenly contrary to how you would define success.

To accomplish your goals, however, you need to know how to set them. You can't simply say, "I want." and expect it to happen. Goal setting is a process that starts with careful consideration of what you want to achieve, and ends with a lot of hard work to actually do it. In between there are some very well defined steps that transcend the specifics of each goal. Knowing these steps will allow you to formulate goals that you can accomplish. Here are our Five Golden Rules of Goal Setting:

The Five Golden Rules

Rule No. 1

Set Goals that Motivate You When you set goals for yourself, it is important that they motivate you: This means making sure it is something that's important to you and there is value in achieving it. If you have little interest in the outcome, or it is irrelevant given the larger picture, then the chances of you putting in the work to make it happen are slim. Motivation is key to achieving goals.

Set goals that relate to the high priorities in your life. Without this type of focus you can end up with far too many goals, leaving you too little time to devote to each one. Goal achievement requires commitment, so to maximize the likelihood of success, you need to feel a sense of urgency and have an "I must do this" attitude. When you don't have this "must do" factor, you risk putting off what you need to do to make the goal a reality. This in turn leaves you feeling disappointed and frustrated with yourself, both of which are de-motivating. And you can end up in a very destructive "I can't do anything or be successful at anything" frame of mind.

Rule No. 2: Set SMART Goals

You have probably heard of "SMART goals" already. But do you always apply the rule? The simple fact is that for any goal to be achieved it must be designed to be SMART. There are many variations on what SMART stands for, but the essence is this - Goals should be:Specific Measurable Attainable Relevant Time Bound

Set Specific Goals

You goal must be clear and well defined. Vague or generalized goals are not achievable because they don't provide sufficient direction. Remember, you need goals to show you the way. How useful would a map of the United States be if there were only state borders marked on it and you were trying to get from Miami to Los Angeles? Do you even know which state you are starting from let alone which one you're headed to? Make it as easy as you can to get where you want to go by defining precisely where it is you want to end up.

Set Measurable Goals

Include precise amounts, dates, etc in your goals so you can measure your degree of success. If your goal is simply defined as "To reduce expenses" how will you know when you are successful? In one month's time if you have a 1% reduction or in two year's time when you have a 10% reduction? Without a way to measure your success you miss out on the celebration that comes with knowing you actually achieved something.

Set Attainable Goals

Make sure that it's possible to achieve the goals you set. If you set a goal that you have no hope of achieving you will only demoralize yourself and erode your confidence.

However, resist the urge to set goals that are too easy. Accomplishing a goal that you didn't have to work very hard for can be an anticlimax at best, and can also make you fear setting future goals that carry a risk of non-achievement. By setting realistic yet challenging goals you hit the balance you need. These are the types of goals that require you to "raise the bar" and they bring the greatest personal satisfaction.

Set Relevant Goals

Goals should be relevant to the direction you want your life and career to take. By keeping goals aligned with this, you'll develop the focus you need to get ahead and do what you want. Set widely scattered and inconsistent goals, and you'll fritter your time - and your life - away.

Set Time-Bound Goals

You goals must have a deadline. This again, is so that you know when to celebrate your success. When you are working on a deadline, your sense of urgency increases and achievement will come that much quicker.

Rule No. 3: Set Goals in Writing

The physical act of writing down a goal makes it real and tangible. You have no excuse for forgetting about it. As you write, use the word "will" instead of "would like to" or "might". For example, "I will reduce my operating expenses by 10% this year." Not, "I would like to reduce my operating expenses by 10% this year." The first goal statement has power and you can "see" yourself reducing expenses, the second lacks passion and gives you an out if you get sidetracked.

Post your goals in visible places to remind yourself everyday of what it is you intend to do. Put them on your walls, desk, computer monitor, bathroom mirror or refrigerator as a constant reminder. You can even post them in the Mind Tools Career Excellence Club forum and share them with other members, for added motivation.

Rule No. 4: Make an Action Plan

This step is often missed in the process of goal setting. You get so focused on the outcome that you forget to plan all of the steps that are needed along the way.

By writing out the individual steps, and then crossing each one off as you complete it, you'll realize that you are making progress towards your ultimate goal. This is especially important if your goal is big and demanding, or long-term. Read our article on Action Plans for more on how to do this.

Rule No. 5: Stick With It!

Remember, goal setting is an ongoing activity not just a means to an end. Build in reminders to keep you on track and remember to review your goals continuously. Your end destination may remain quite similar over the long term but the action plan you set for yourself along the way can change significantly. Make sure the relevance, value, and necessity remain high.

Key Points

Goal setting is much more than simply saying you want something to happen. Unless you clearly define exactly what you want and understand why you want it the first place, your odds of success are considerably reduced. By following the Five Golden Rules of Goal Setting you can set goals with confidence and enjoy the satisfaction that comes along with knowing you achieved what you set out to do. What will you decide to accomplish today?

BACKWARD GOAL-SETTING

If your goal is to become an account executive within the next five years, where do you start your planning process? Or if your team needs to redesign the company's organizational structure, where do you begin?

In planning, most of us would usually start building our plan from start to finish. What do you have to do first, second, third, and so on? What milestones do you have to reach before you can continue on with the next step in the plan? By what date does each step need to be completed?

This is a solid form of Personal Goal Setting that works very well. When combined with the Golden Rules of Goal Setting, you have a motivating formula that can help you actively move yourself forward.

A New Approach

However, there's another, simple but lesser-used method of goal setting that can be equally as powerful. It can show you other ways to achieve the same result, and it can help you deal with the necessary unknowns of goal setting that can so often cause you to give up on your plan entirely.

It's called backward planning, backward goal-setting, or backward design, and it's used quite often in education and training. The idea is to start with your ultimate objective, your end goal, and then work backward from there to develop your plan. By starting at the end and looking back, you can mentally prepare yourself for success, map out the specific milestones you need to reach, and identify where in your plan you have to be particularly energetic or creative to achieve the desired results.

It's much like a good presentation, when the presenter tells you where he's headed right at the beginning. Then, as the presentation unfolds, it's easy for you to follow the concepts and think critically about what's being said. If you have to figure out the main points as they come, your energy is often used up by just trying to keep up.

The Backward Planning Process

Here's how it works:

1. Write down your ultimate goal. What specifically do you want to achieve, and by what date? Example: "By January 1, 2013, I will be the key accounts director for Crunchy Chips International."
2. Then ask yourself what milestone you need to accomplish just before that, in order to achieve your ultimate goal. What specifically do you have to do, and by when, so that you're in a position to reach your final objective? Example: "By September 30, 2011, I will successfully complete the executive training program offered by Crunchy Chips International."
3. Then work backward some more. What do you need to complete before that second-to-last goal? Example: "By March 1, 2011, I will submit my application for the executive training program, outlining my successes as a key accounts manager, and I will be accepted into the program."
4. Work back again. What do you need to do to make sure the previous goal is reached? Example: "By January 1, 2011, I will complete my second year as a key accounts manager with Crunchy Chips International, and I will earn the prestigious Key Accounts Manager of the Year award."
5. Continue to work back, in the same way, until you identify the very first milestone that you need to accomplish. Example: "By January 1, 2010, I will complete my first year as a key accounts manager with

Crunchy Chips International, and I will be rewarded for my performance by gaining responsibility for clients purchasing over $10 million per year."

"By January 1, 2009, I will be promoted to key accounts manager with Crunchy Chips International, and I will have responsibility for clients purchasing over $1 million per quarter." When you read a backward plan, it doesn't look much different from a traditional forward plan.

However, creating a backward plan is VERY different. You need to force yourself to think from a completely new perspective, to help you see things that you might miss if you use a traditional chronological process.

This can also help you avoid spending time on unnecessary or unproductive activities along the way. Furthermore, it highlights points of tension within the plan, showing where you'll need to be particularly creative to make the next step successfully.

KEY POINTS

On the surface, backward planning doesn't seem much different from traditional goal-setting processes. You start with a basic vision, and then you ask yourself what needs to be done to achieve that vision. You can read your plan from the beginning to the end, or from the end back to the beginning.

Backward planning, however, is more than reversing the direction of your traditional plan. It's about adopting a different perspective and, perhaps, identifying different milestones as a result. It's a great supplement to traditional planning, and it gives you a much fuller appreciation for what it may take to achieve success. After all, the more alternatives you have, the better your final plan will likely be.

LEVERAGE

To lift a heavy object, you have a choice: use leverage or not. You can try to lift the object directly – risking injury – or you can use a lever, such as a jack or a long plank of wood, to transfer some of the weight, and then lift the object that way. Which approach is wiser? Will you succeed without using leverage? Maybe. But you can lift so much more with leverage, and do it so much more easily!

So what has this got to do with your life and career?

The answer is "a lot". By applying the concept of leverage to business and career success, you can, with a little thought, accomplish very much more than you can without it.

Without leverage, you may work very hard, but your rewards are limited by the hours you put in. With leverage, you can break this connection and, in time, achieve very much more.

LEVERS OF SUCCESS

So how can you apply leverage to your career? And how can you achieve much more, while – if you choose to – reducing the number of hours that you work?

To do this, you'll need to learn how to use the leverage of:-

- Time.
- Resources.
- Knowledge and education.
- Technology.

Time Leverage

Using the leverage of time is the most fundamental strategy for success. There are only so many hours in a day that you can work. If you use only your own time, you can achieve only so much. If you leverage other people's time, you can increase productivity to an extraordinary extent.

To leverage YOUR OWN time:

- Practice effective time management. Eliminate unnecessary activities, and focus your effort on the things that really matter.
- As part of this, learn how to prioritize, so that you focus your energy on the activities that give the greatest return for the time invested.
- Use goal setting to think about what matters to you in the long term, set clear targets, and motivate yourself to achieve those targets.

To leverage other people's time:

- Learn how to delegate work to other people.
- Train and empower others.
- Bring in experts and consultants to cover skill or knowledge gaps.
- Outsource non-core tasks to people with the experience to do them more efficiently.

Providing that you do things properly, the time and money that you invest in leveraging other people's time is usually well spent. Remember, though, that you'll almost always have to "pay" up front in some way in order to reap the longer-term benefits of using leverage.

Resource Leverage

You can also exert leverage by getting the most from your assets, and taking full advantage of your personal strengths.

You have a wide range of skills, talents, experiences, thoughts, and ideas. These can, and should, be used in the best combination. What relevant skills and strengths do you have that others don't? How can you use these to best effect, and how can you improve them so that they're truly remarkable? What relevant assets do you have that others don't? Can you use these to create leverage? Do you have connections that others don't have? Or financial resources? Or some other asset that you can use to greater effect? A good way of thinking about this is to conduct a personal SWOT analysis, focusing on identifying strengths and assets, and expanding from these to identify the opportunities they give you.

Knowledge and Education Leverage

Another significant lever of success is applied knowledge. Combined with education and action, this can generate tremendous leverage. Learning by experience is slow and painful. If you can find more formal ways of learning, you'll progress much more quickly. What's more, if you select a good course, you'll have a solid foundation to your knowledge, and one that doesn't have high-risk gaps. This is why people working in life-or-death areas need long and thorough training. After all, would you want to be operated on by an unqualified surgeon?

While few of us operate in quite such immediately critical areas, by determining what you need to know, and then acquiring that knowledge, you can avoid many years of slow, painful trial and error learning.

In the same way, it's inefficient if many people in an organization have to learn how to do their work by trial and error. A much better way is for organizations to capture the knowledge gained by the first few in some way and pass it on to others. This is the core of the "knowledge management" concept. Premium Members of the Career Excellence Club can hear more about this in our Book Insight on The Complete Idiot's Guide to Knowledge Management.

The keys to successfully leveraging knowledge and education are: firstly, knowing what you need to learn; secondly knowing to what level you need to learn it; thirdly, being very focused and selective in your choices; and fourthly, in taking the time to earn the qualifications you need. Even then, having more education or more knowledge isn't necessarily a point of leverage. These become advantages only when they can be directly applied to your career goals and aspirations—and when they're used actively and intelligently to do something useful.

Technology Leverage

Finding technology leverage is all about thinking about how you work, and using technology to automate as much of this as you can. At a simple level, you might find that all you need to keep you in touch with home and work is a laptop computer. Alternatively, a personal digital assistant can help you maintain a single, convenient, properly-backed-up time management system. Cell phones that access email and browse the web are handy tools for making the best of your downtime during working hours or while traveling. If you're a slow typist, voice recognition software can help you dictate documents and save time. Tools like Google Desktop Search can help you manage and find documents in such a way that you no longer need to file digital documents. And Google itself provides a great, quick way of finding relevant information online. At a more sophisticated level, you may find that you can use simple desktop databases like Microsoft Access to automate simple work processes. If you do a lot of routine data processing you can find that this saves you a great deal of time. More than this, you only need to set up a process once with a tool like this - afterwards the process will be executed the same way each time, by whomever initiates the process

Businesses can choose from a wide array of software solutions. Some of these can automate or simplify tasks that are otherwise very time-consuming. Customer relationship management databases can bring tremendous benefits for sales and customer service organizations, as can point-of-sale inventory systems for organizations that need to track and manage inventory. Websites and web-based catalogs can give clients easy access to up-to-date product information, and help them place orders simply and easily. And blogs and email-based newsletters help people stay in contact with thousands of people quickly and easily. All of these use technology to provide tremendous leverage.

KEY POINTS

Using leverage is the art and science of getting much more done with the same, or less, effort. At a simple level, this can free up your time to concentrate on things with the highest priority. At a more sophisticated level, it helps you achieve at a much higher level. When you invest time and resources to leverage technology - as well as to leverage time, resources, and knowledge —you have a recipe for unprecedented success. Use what you and others have to your advantage, and see how far it will take you.

4

Marriage

MARRIAGE AND FAMILY: CHOOSING LIFE PARTNER

MEANING OF MARRIAGE

Marriage is a common term that we come crossways in our daily life. Have you ever thought of it seriously? What is the meaning of marriage? Is it a mere relationship flanked by a man and woman to live together and have children? Is it a man-made institution? When did such an institution come into subsistence?

It will be really motivating to ponder in excess of the answers to the above questions – much more so in relation to the meaning, scope, purpose and history of marriage. The term 'marriage' has dissimilar meanings and connotations for dissimilar people.

To some, marriage is a relationship flanked by men and women for the propagation of the human species. Some people take it as a license for sex. Yet another group considers marriage as a means for companionship. Marriage comprises all these views and much more. It is a very intricate institution which can not be defined in a sentence or two. Several sociologists and philosophers have given differing views in relation to the meaning and scope of marriage.

Marriage is not an integral part of human nature, but it is a man-made custom or institution which was present even in prehistoric times. It is not a natural relationship but a conscious commitment flanked by a man and a woman. With the advancement of civilization marriage became a social function with religious and legal sanction. As stated earlier, marriage has dissimilar meanings and connotations. Let us look at some of them.

Marriage as a Relationship

Marriage is one of the deepest and mainly fulfilling of human relationships. It has lived in varying shapes throughout the history of man, responding to the fundamental needs and social aspirations of each generation.

Marriage as an Institution

Marriage is a union of man and woman their bodies, minds and souls, emotions and desires. The essence of this union is love. Marriage is measured to be a primary relationship, because it is a personal relationship flanked by the partners. Lin Yutan gives a beautiful explanation of marriage. He says "woman is water and man is clay and the clay holds the water and gives it substance in which water moves and lives and has its full being".

The Biological Aspect of Marriage

The animal mates, but man marries. Looking from the biological point of view, we can say that mating is a biological matter, while marriage is a social affair. The sex instinct is one of the vital instincts for reproduction. In the case of man, though, from the beginning of society, this instinct has been subjected to separate regulations and control.

Marriage may be defined as the mechanism of regulation and control of biological reproduction. In the animal kingdom there are no definite rules for mating because they have no society and social norms. But in human society there are several kinds of social controls, religious limitations and even laws of marriage. For human beings, the sex relations are allowed within sure permitted limits. Marriage is a holy sanctifying, life extensive, comprehensive, loving union of a man and a woman, leading to the procreation of children.

The Social Characteristics of Marriage

We have seen that there are sure biological characteristics of human marriage, while there are more social characteristics in it. These social characteristics are mainly significant in human society. The vital factor in marriage is love. True love flanked by the man and woman in marriage is a complete giving of one to another. There should be a union of the souls and minds. There should be a union of emotions and wills. All these imply complete, exclusive and life extensive giving of ones self to the other.

The Psychological Aspect of Marriage

Marriage is an integral part of human subsistence. In marriage man's vital longing for communication and completeness finds its natural culmination. Human beings find their fulfillment as men and women. Marriage is the means through which two persons seeking a more perfect life provide themselves totally to each other.

In a marriage the male and female compliment each other. It fills the physical and psychological void which exists in human being, if they are alone. Marriage makes possible the satisfaction of the strong and natural tendency of male and female to be together in the union for which nature has intended them.

The Legal Aspect of Marriage

The union flanked by man and woman necessity is legitimate. So marriage should get legal, social and civil acceptance. Legal sanction of marriage is based on prevailing social norms and customs. It varies from one society to another. The marriage can be legally contracted only through those who are able to perform the vital act of marriage. The legal minimum age for marriage in India is 18 years for girls and 21 years for boys.

THE FUNCTIONS AND PURPOSES OF MARRIAGE

Have you ever thought of the question why a person should get married? The first answer that comes to your mind may be for the propagation of the human species. Let us try to find the answer.

Marriage for Union and Procreation

What is the purpose of marriage? If it is only for procreation, marriage is not a necessity. Of course, one of the significant purposes of marriage is union and procreation. The union in marriage is a medium of physical, psychological and spiritual communication of love and commitment of self.

Marriage for Sex

The natural end of sexual relationship is conception of children. Hence the procreation of children is an essential purpose of marriage. An equally significant purpose is the union itself, the mutual love, pleasure and happiness of the husband and wife. Marriage, so, fosters mutual love and attachment. It provides for the legitimate expression of sexual satisfaction.

Marriage for Companionship and Friendship

The mainly significant need of the human being is the intimacy in living with and making a commitment to another human being. What is friendship? According to Jennet Kid, "Friendship is having a privileged position in someone else's life and giving them a privileged position in our own. It is sharing ourselves with those we like". Friendship is the cornerstone of marriage which lasts even when the sexual desire is in excess of. It remnants even after the children are grown up and settled. It only deepens with years. This is what is meant through companionship or friendship. It enriches man and woman through rising unselfishness and through deepening his/her capability for love, and sacrifice. Hence marriage is love, it is sex, it is family, but ultimately and essentially it is companionship or friendship.

Marriage for Socialization

Marriage is a means through which the socialization and growth of a person reach its fulfillment. It provides numerous opportunities to bring in relation to the security, cooperation, and love. Another purpose of marriage is to create a family to give the natural environment in which a person can realize himself/herself and reach out to others in an attitude of dedication and service. It gives society a firm base, and children a stable environment to grow.

Marriage for Matured Relationship

Still another purpose of marriage is attainment of maturity through the establishment of relationships in marriage. Rearing and educating the children also is a purpose of marriage. The parent-child relationship is an intimate relationship.

Some Practical Purposes or Utilitarian Characteristics of Marriage

- It ensures security to women who have to undergo extensive periods of pregnancy.
- It provides security to the offspring.
- It ensures the health life which gives stability to society.
- It makes society more cohesive through relationships.
- It simplifies blood relationships.

THE HISTORICAL DEVELOPMENT OF MARRIAGE

It is motivating to study the historical development of marriage. When did it commence? How did it attain its present structure and status? It is not easy to get the facts in relation to the marriage in the prehistoric era. We have to depend upon the studies of anthropologists and other accessible details of folklore and traditions. Even though the religious books mention marriage, we cannot conclude that marriage came into subsistence with the beginning of human life on earth. In the pre-historic ages people lived like animals and there were no social norms and rules to guide them. But slowly due to his superiority in excess of the animals, man began to formulate some social patterns of behaviour.

In the early stages of human life, sexual life and procreation was a natural procedure. The earth was not owned through anybody. Nobody bothered to own the land. But slowly man began to cultivate the land and came to know in relation to the productivity of the soil. This knowledge motivated him to own the land.

This also led to several disputes and fights for the ownership of land. Finally they reached on agreement with regard to some kind of control or laws in relation to the ownership of land. This is apparent from social contract theory. The desire of man to own the land, which is productive, led him to the thought of owning any thing that is productive. Knowing that the woman can procreate children, men tried to own more women and thereby get more and more children. This also ended the conflicts and fights. Slowly restrictions and regulations came in procuring women. This finally culminated in the form of a social institution of marriage with all its laws and regulations and customs.

THE FAMILY

Presently like marriage, family is also a very familiar term. We all come from families. We see families approximately us. Family is an integral part of our life. Can we answer these questions: What is a family? What is the need for a family? Who constitute the family? Is it a mere union of parents and children? What is the relationship flanked by the members in a family? Is the family a human invention or a natural development? Is it found in the middle of animals also? What are the shapes of family in dissimilar ages and dissimilar cultures?

The Need for a Family

The family was a necessary precondition for human development, for without it human subsistence would not have been possible. The human infant is born helpless. It has a potential for physical and mental development. But it requires years to achieve its maturity. In the case of animals, their young ones are able to take care of themselves shortly after birth. But for human beings the extensive period of pregnancy and the prolonged helplessness in infancy and childhood, needs a extensive association of parents. This has resulted in the formation of a family.

The Development of the Family

The family as a group of parents and offspring's lived even before the appearance of man upon this planet. Family like association is seen in the middle of birds and higher mammals. The chimpanzee is a highly social creature which lives in family groups. The family life of apes and human beings can be compared. There is a selection of a mate, interaction flanked by male and female, stages of control in the middle of father, mother and children and a child is taken care of primarily through mother.

There is a great deal of variation flanked by the family life of apes, mammals and birds and that of man. In the former case, the nature of family life throughout the world is the same in any given species. But in the case of man, family behaviour varies greatly from one society to another. In the animal family, the behaviour of the members is motivated through instinct, whereas in the human family it is motivated through culture.

The animal family is mainly biological in nature, while in the human family its structure and functions are shaped through culture. According to several sociologists, man in course of his development might have urbanized one or another family form as determined through situational and historical factors. In short we can say that geographic environment, economic circumstances and culture rather than biological factors were the determinants of human family patterns. In man, the family is not only a biological group it is first and foremost a social institution.

The Definition of Family

We have seen that the structure and function of the family varies from place to place. So it is hard to provide a definition of the term 'family'. Though, the definition of the family necessity contain that which is common to the great variety of human groups to which the term 'family' has been applied. There are sure features that are common to the human family in all times and in all places that differentiate the family from other social groups. According to Ernest W Burgess and Harvey. J. Lock, these features are:

- The family is composed of persons united through the ties of marriage, blood or adoption. The bond flanked by husband and wife is that of marriage, and the relationship flanked by parents and children is usually that of blood and sometimes that of adoption.
- The members of the family typically live together under one roof and constitute a single household. Sometimes, as in the past, the household is large consisting of three or four or even five generations. Today the household is small, consisting of the husband and wife with or without one or two children.
- The family is a unity of interacting and intercommunicating persons. They play the roles of husband and wife, father and mother, son and daughter, brother and sister. These roles are defined through the community.
- The family maintains a common culture derived mainly from the general culture. Usually this culture is the outcome of the merging of the two cultures of the husband and wife.

On the foundation of the above features the family may be defined "as a group of persons united through the ties of marriage, blood or adoption, constituting a single household; interacting and communicating with each other in their respective social roles as husband and wife, mother and father, son and daughter, brother and sister and creating and maintaining a common culture". A family is a community of people living together in an environment which is a centre of healing, a place where one can live, where one can admit one's frustrations, stupidities and anger to people who do not retaliate.

It is in the school of everyday family life, with all masks dropped and hypocrisies exposed, that man, woman and child acquire the potential to know what it is to be truly human, and fully human people have tried alternative arrangements for a family. But none of them are as efficient as a family. Children thrive best when they have a father and mother to love and protect them in a home. The parents also need the life extensive commitment in which love is proved.

The Indian Family

The family system all in excess of the world are changing today and family relations are becoming weaker and weaker day through day. The power of the family in excess of the children is weakening due to the great social, political and economic changes. In the midst of these changes, India still has a time tested heritage of stable family structure. It is still the vital unit of our society and the medium of cultural transmission. The family still holds its solidarity and plays an significant role in the formation of values in our society. In spite of the solidarity in the Indian families, a salient social transformation is taking place in the Indian families also. With the empowerment of women and the women getting more and more educated, the traditional family patterns are slowly disappearing. Modernization has its roots in westernization and it has questioned the traditional family values and structures. Parental power is also weakening day through day. The religious and moral values present in the family are giving way to secular and pragmatic values.

KINDS OF FAMILY PATTERN

The three chief historical stages in the development of the family are:

- The large patriarchal family feature of ancient society;
- The small patriarchal family which had its origin in the medieval period; and
- The modern democratic or nuclear family which is the product of the industrial revolution and the economic and social changes that followed.

The large patriarchal family was prevalent in China, India and Japan. Here the senior male member was the head of the family. He lived with his children and grand children which came up to four or five generations. The patriarch, the head of the family, had supreme power in excess of all other members of the family. This kind of family was a result of the agricultural and rustic mode of subsistence. The small patriarchal family was the second stage in the development of the family structure.

It consisted of husband and wife and children and one or two grandparents and one or two unmarried brothers and sisters of the husband and wife. This kind of family was mostly seen in urban areas where the members used to work in industries. Here also the senior male member had the supreme power in excess of the whole family. The democratic or nuclear family consists of only the husband and wife and the children. This kind of family system is more prevalent in modern society. Here the husband and wife share more responsibility and are more free. They can take their own decisions. Children can also join in the decision making constant with the advancement in age. In India we find three kinds of family structures which are approximately identical with the historical family patterns. They are:

- The large joint family;
- The nuclear family; and
- The stem or extended family

Joint Family

The large joint family is approximately like the large patriarchal family where three or four generations of parents and off springs live together. This is mostly seen in rural areas. These families are mostly agricultural families.

- Advantages of the Joint Family: The joint family assures shelter for the aged and sick, social security for the unemployed and support for the young couple. Children grow up in an atmosphere of security and affection. The newly married couple gets training in family life and child care. The sons get the training from their father, uncle and grandfathers. The daughters get it from their mothers, aunts and grandmothers. In a joint family, the wisdom and experience of the elders are shared. The joint family has its own codes of behaviour and its own values which are transmitted from one generation to the other generation. On the whole, we can say that the joint family provides an umbrella of support which covers financial loss, social security and even provides informal counseling.
- Disadvantages of the Joint Family: In the joint family the supreme power is vested in the senior male member. So the whole life of the family goes according to his efficiency and attitudes. The junior members may not take up any responsibility and initiative. There is no freedom for the individuals, especially women. The head of the family may not be able to adjust to the social changes that are taking place outside the family. There will be a perpetuation of old customs and values.

Nuclear Family

In a nuclear family, the husband and wife live with their children. This is mostly seen in urban areas. Both the husband and wife may be earning members in such families.

- Advantages: The husband and wife have the full freedom to act according to their own ideas. There is more financial security and individual freedom. This kind of families can easily adapt to social changes. There is more responsibility and initiative for all the members.
- Disadvantages: There is nobody to help and guide the members, especially when some disagreement arises. The practical wisdom and emotional security offered through the joint family is lacking in a nuclear family. There is nobody to look after the children. Children

miss the protection and affection of grand parents. Working mothers are forced to leave the children either with the servants or in a day care centre.

The Stem or Extended Family

This is midway flanked by the joint family and the nuclear family. The husband, wife and children live with one of their grand parents.

- Advantages: This kind of family has all the advantages of the joint family and the nuclear family, provided the grandparents do not dominate. Children are looked after properly. Grandparents also may not feel the loneliness and will be happy with their children and grandchildren. Parents can provide all the security and guidance to their son and daughter, who are newly married.
- Disadvantages: If the parents who live with their son/daughter are too dominating, the young couple may loose their freedom and individuality. There are chances of troubles with in-laws. For growth in marriage, it is desirable that the couple live on their own. Each system of family has its own advantages and disadvantages. But if the members are cooperative and have concern for each other, the disadvantages can be reduced to the minimum. The newly married couple who start their family life should have the freedom and initiative of a nuclear family and the emotional security and practical wisdom of a joint family. They should have the feeling of the 'home' where one has full relaxation and recreation and can live without masks.

Functions of the Family

The family is the vital unit of society. Today's family faces lot of troubles due to the social changes. In spite of the troubles in the family, it continues to exist because it meets the needs of children, adults and society at large. The family:

- Provides for the reproduction of the race;
- Passes on the cultural heritage of the group;
- Provides physical security, protection and the material opportunities for living and growth;
- Meets the deep emotional needs of both children and adults and provides for their social, emotional, intellectual and spiritual development;
- Develops in its members socially desirable character traits and acceptable moral standards;
- Develops an orderly system of living in the middle of its members with provision for eating, sleeping, school, work, etc., and

- Develops sound relationships in the middle of members of the family and flanked by them and their neighbors.

The family thus meets the vital physiological and psychological needs of its members.

The Social Role of the Family

Separately from the above functions the family has a social role also. The family, being the vital unit of society, has a vital and organic link with society. It is from the family that the citizens come and it is within the family that they find the first school of social virtues. The family is the bridge flanked by the child and society. In the family, the child learns mutual respect, concern for others, generous service, deep solidarity, personal responsibility etc. These are the vital steps of social life. The family is thus the mainly effective means of humanizing and personalizing society. The family is the custodian and transmitter of values.

Society should also respect and foster the family. The family is the very first cell in the social structure and the very first social union, which is indispensable. The so-described urbanized countries have now realized the fact that their broken families are heading towards a great social disaster so they are exhorting the parents to strengthen the family ties. "Marriage and family are a union for which there is no substitute. Nothing can take their place. You can even say: as the families are, so will mankind be".

CHOOSING A PARTNER

Now that you have learned so much in relation to the marriage and family, you will be in a position to answer the question "How will you choose your partner for life"? Marriage is a life-extensive commitment and relationship and hence the selection of a partner is very significant. The choice of an ideal partner is one of the vital requirements for a successful marriage. Suppose you are going for a journey. You would like to have a companion who is agreeable to you in all respects. So what in relation to your companion, who is to travel with you through your whole life, till death? Marriage is one of the three great events in life beside with birth and death. Birth presently happens and death is beyond our control. Marriage is though something which can be decided. We can decide whom to marry and when to marry. There is a common saying in the middle of the Arabs " If you want to select a horse, select from in the middle of hundred horses. If it is a friend it should be one in the middle of thousand. But if it is a wife, it is should be from thousands". This implies the importance of choosing a partner in marriage. It should not be a casual or careless selection. It should be a extensive thought out procedure.

Very few people are inclined to analyze the factors that are to be looked into in selecting a partner. There is a common belief that marriages are made in heaven or it is fate, luck or providence. But this issue of choice of spouse is not something to be totally left to fate or luck. A scientific approach is necessary in the selection. There are sure determinants for mate selection. The patterns of arrangement for the selection of the mate differ from culture to culture.

In sure cultures marriage is primarily a social and economic arrangement flanked by two families. Here the choice of the mate is still the responsibility of the parents. Parents select the bride or bridegroom for their children. The young couple has little choice in this matter. They may not have the opportunity to become acquainted before the wedding. This kind of marriage is recognized as arranged marriage.

Arranged marriages are now giving way to selection through the young people themselves. Both procedures have their advantages and disadvantages. Mate selection, when arranged through parents, stressed social and economic thoughts. They minimized or even ignored the sentiments of love and interpersonal relations of the young people. The youngsters make the choice on the foundation of compatibility and personal attraction.

They may not conform to parental standards like caste, religion, economic security etc. The best pattern is to let the young people find out heir own mates with the guidance and consent of the parents. This pattern is recognized as guided choice. Anyway, parents should guide their children without showing too much worry or haste. Allow the young people to decide on their partners as far as they can. But in every case, their elders should be the controlling factor. At the same time, parents should not provide the impression that they have a readymade decision in their minds. Youngsters on their part should rely on their parents, recognizing their experience and their sincere desire for the good of their children.

As said earlier, marriage is a life-extensive relationship and commitment. There are sure qualities or social expectations that are commonly appreciated in a partner. There should be compatibility flanked by the partners in maturity, health, behaviour patterns or character, caste and religion, economic status, education and intelligence, attitudes, and values or the vital orientation to life etc. Let us consider them one through one.

Maturity

Under 'maturity' come physical maturity, emotional maturity and social and intellectual maturity. Maturity is a term which comprises several things. When we say that the partner should have the maturity for a married life, it means the ability to take up the full responsibility of a family. Age is one significant factor that comes under physical maturity. According to the Indian Marriage Act, the minimum age for marriage for woman is 18 and for man are 21. But from practical experience it is seem that the ideal age for woman is 21 to 24 and for man is 25 to 30. Through that time they are physically mature and emotionally stable. If the man or woman lacks physical maturity he/she may find it hard to make marital adjustments and to take up the responsibility as husband and wife and as parents.

What should be the age variation flanked by the partners? Who should be older? The social expectation is that the man should be older and there should not be much variation of age flanked by the partners. But for so several practical reasons, it may not be possible to strictly adhere to this. Still, it is always good to stick on to these expectations to avoid further complications later on. One of the reasons for marital breakdown is the great or wide disparity of age flanked by husband and wife.

Emotional Maturity

It consists of the ability to control one's emotions and passions. Emotional maturity helps to develop self-restraint and the attitude of self sacrifice. These are the essential requisites for a happy married life. Emotional maturity can be tested only at the times of crisis. He will adapt his course of action, overcome obstacles and accept the inevitable with grace and calmness. To an immature person, any frustration may bring in relation to the temper tantrums. An emotionally matured person is able to avoid frequent bad mood and has the ability to establish and uphold personal relationships. He is able to endure normal discomforts and disappointments and to overcome suspicion and jealousy. In short, he has the ability to provide and receive that is ability to love.

Empathy is another feature of an emotionally matured persons. It is ability to perceive the feelings of others. It involves the willingness to recognize the needs of others and to assume the responsibility of meeting them. Getting married means taking up a responsibility for a life time. The partners have to meet each other's needs, bring up children, support the family financially and look after the members of the family. Life extensive commitment is one characteristic of the responsibility of marriage. It is related to the stability that comes with maturity. A clear sign of emotional maturity is the ability to reflect before speaking and the readiness to talk with others. This is highly essential in husband/wife relationship. An emotionally mature person has, first of all, insight and foresight in his thinking. He can evaluate himself as well as the world approximately him in a realistic manner. He can also face the facts of life realistically and expect the results of his action. He develops a sense of independence. He is able to make his own judgments and decisions.

The maturity which marriage requires is not the attainment of one day. Emotional maturity can be achieved only through the satisfaction of the several needs of the child at each stage of development. Proper disciplining of emotions is necessary throughout the childhood for attaining emotional maturity, which results in self-confidence, self-control and affect ional maturity. What is affect ional maturity? It may be an unfamiliar term for you. It is the capability to relate with people of dissimilar groups. The infant is at first interested in himself/ herself or his/her mother. This interest is then shifted to playmates of his own sex throughout school age and to the opposite sex throughout adolescence. When the person comes to maturity, the affect ional maturity is towards his partner. One who is ready for marriage should have this affect ional maturity so as to develop a good husband wife relationship. This is vital requirement for marital harmony.

Social Maturity

Social maturity is apparent when one can relate oneself to others in a selfless and responsible way. One should not think only in terms of immediate wish fulfillments or satisfaction or personal desires. The main signs of social maturity are respect for others, honesty, frankness, courage and the ability to give whatever a family needs. An emotionally mature person need not necessarily be socially mature if he has not experienced a social life. Social maturity comes from the fulfillment of one's quota of premarital living. Every boy and girl should have an experience of social life before marriage. Usually after completing this education, youngsters are employed and they have a 'free' time to assume the responsibilities of life and mingle with people. Throughout this time they may come crossways persons of other sex, whom they may choose as their partners in life. This is a time of fantasy. Any way, this free time before marriage is essential for acquiring social maturity.

After exploring for a partner, they finally settle down, and are ready for marriage. One feature of social maturity is one's willingness to disregard unknown potential marriage partners in their fantasy. They will be ready to commit themselves to build up a relationship with a scrupulous person. It is also significant to be self-governing for a while. Having presently become free from parents, it is too early to take one the bonds of matrimony. Spare time can be used as one may deem fit. Jobs can be changed. Travels can be made according to the availability of money and time. Nobody else's wishes have to be consulted, nor moods catered to. Since they were more closely supervised than boys throughout childhood and adolescence, they have yet to achieve a full sense of personal identity. Usually several girls in India, especially in the rural areas, do not get this free time for socialization and acquiring social maturity. This is because they are married immediately after their schooling or graduation. Mainly of them are not permitted to work outside their homes. This is a real drawback as far as marriage is concerned.

Intellectual Maturity

Intellectual maturity lies in the ability to understand persons, events, situations and troubles. It is the capability to formulate one's ideas, opinions and judgments without depending on others. For developing intellectual maturity one necessity has a sure stage of education according to his/her social status. One necessity is able to provide a meaning and purpose to his/her life, if he/she is intellectually mature. As we grow up, we slowly learn to understand and evaluate ourselves better. We come to know of our own strengths and weaknesses, our abilities and disabilities. We gain insight into our feelings, thinking and behaviour. We also develop foresight. We learn to face realistically the facts of life and to expect the results of our action. We learn to foresee the possible consequences of our behaviour. We no longer permit our desires to dominate totally our thinking or action.

Mature thinking and feeling express themselves in mature action and behaviour. We show control and flexibility in our behaviour. We are neither rigid nor compulsive but adapt our actions according to the situation. We can accept power and discipline as well as responsibility and power. We learn to co-operate with the people and make the necessary adjustments and version in life. An individual should show these signs of intellectual maturity before he/she plans to marry. It is essential for promotion of desirable personality development and adjustments in marriage. Self realization is essential to understand the behaviour of others. "Know thyself" is an significant principle in mental health. For good marital adjustments, knowledge of self is an essential factor. Only an intellectually matured person can know the limitations of others. A person who aims at an unattainable goal meets with frustrations. Intellectual maturity is essential for having 'frustration tolerance'. The married partners need high frustration tolerance to face problem situations wisely and to make wise decisions.

Now we have seen how significant maturity in selecting a mate is. Only a person having all the above kinds of maturation will be a good partner in married life. Often it is not easy to evaluate correctly, the herself in the best high, hiding the negative characteristics on his/her personality. So in mate selection, it is necessary to go beyond the external appearance and behaviour. You will have to closely observe the behaviour of the person in dissimilar situation and circumstances. How a person acts in a crisis or a critical situation will show his/her real self and maturity.

Health and Physical Structure

A happy married life is very closely related to the health of the partners. So this is an significant factor to be looked into in selecting a mate. The parents should be healthy to have children of good health. In some western countries, the partners have to produce a medical certificate before marriage. This is ideal, especially to eliminate people with hereditary diseases. It is advisable not to have marriage relations flanked by person having a close blood relationship. The children born from such couples may have hereditary diseases of both the families. It is also necessary to look out for the RH factor of the blood. If the husband and wife have dissimilar RH groups i.e. positive and negative, it may affect the children. So care should be taken to see that the partners have compatible blood groups. HIV testing also should be done before marriage, wherever scope for doubt exists, to avoid future troubles and breakdowns. The attendance of a health defect in a person may not prevent one from marrying, but it is necessary that both know in relation to it and are ready to face its consequences together.

Physical structure is another factor in the selection of mate. There should not be much disparity in physical structure (height, weight etc. color, and general appearance). Though, physical attraction or physical expectation in the Indian context is that the husband should have a better structure than the wife. Concerning general beauty, the usual practice is that woman's beauty is more looked into than a man's. In a man "manliness" is the usual criteria that are looked for. Physical appearance and beauty are relative terms and vary according to cultures. This is not a very significant factor in selecting a mate. Anyway, beauty should not be the main criteria for selection. There is a Chinese proverb "Marrying a woman only on the foundation of her beauty is like buying a building merely through looking at the outside painting". Young people provide too much importance to this criteria which is not wise.

Behaviour Patterns or Character and Conduct

If you want to enroll for any course, you need a character or conduct certificate. In order to enter into a profession also, very often you need a conduct certificate, but to get married do you produce any conduct certificate? Unfortunately several persons do not provide enough importance to this aspect, which is one of the mainly significant criteria for mate selection. Even if your partner has all the other physical qualities, economic and social status and education, if his character is not satisfactory, married life will be a real hell. At the same time even if there is disparity in age, caste, religion, education etc. and the partner has good character the marriage will still work. Character in married life means good will, sense of humor, the ability to love, honesty, responsibility, concern for others, provide and take attitudes, cheerfulness, discipline, faith in God, fidelity openness, values in life, etc. The couple has a extensive life before them to spend together. Much of their happiness will depend upon their ability to share, to agree and to work together. All this presuppose in them the attendance of strong similarities with regard to vital concepts of life. There is no doubt that a sure amount of differences can be useful to enrich the partners. But ultimately the person who shows more similarities should be measured as the best make.

According to the perspectives in psychology, a person's character is shaped throughout his/her early years of life. So his/her family is mainly responsible for character formation. As in the case of maturity, it is very hard to assess the character of a person with whom we do not have an intimate personal relationship. In choosing a partner this is not always possible. His/her family relationships and the life of the family can be taken as one of the significant sources for assessing the character. Persons brought up in unhappy, insecure, and broken families may have several personality disorders and peculiar behaviour patterns. Some parents consider marriage as a solution for the behaviour troubles of their children, especially sons. A drunkard, a drug addict or a criminal may marry a poor girl having low socio-economic status. The poor girl is supposed to 'change the prodigal son'. Usually this ends up with a marital breakdown and lot of misery for the girl. The innocent girl becomes a scapegoat.

Caste and Religion

When the selection of the bride or bridegroom is made through parents they insist on caste and religion. But when the selection comes to youngster, we come crossways several inter-caste and inter-religious marriages. Several of them are successful. These kinds of marriages are good for social and national integration – and harmony in the middle of dissimilar castes and religions. But there are several practical troubles of adjustments in such marriages. In India marriages are flanked by two families. The families may not accept such marriages and interfere with each one's customs and life-style. When two persons from two dissimilar castes or religions marry, they necessity be willing to accept and respect each other's faith and values. Usually the problem comes with the birth of a child. Both the parents will consciously or unconsciously try to practice his/her own customs and values which result in a disagreement for the children. If both parents are strictly secular, such troubles can be eliminated to some extent. It is safer to choose a mate from the same caste, religion and culture. Even in the urbanized countries, where mate selection is done mostly through couples, preferences are for mates from the same race, religion and culture.

Social and Economic Status

The partners should have as far as possible, the same economic and social status. The family should have a stable income to pull on. Today dowry has become a great social problem in India. Men are demanding vast amounts of money from the wife's home. Several bride burnings are due to the troubles of dowry. Of course, the girl should have an equal share of her father's wealth. Usually this share is handed in excess of to the boy at the time of marriage. In several cases, the wife has no say in relation to the dowry money, and it is measured as the property of the husband and his family. In several marriages, dowry is the main criteria for selecting the wife. This is very unfortunate. The partners should have the financial stability to establish and run a family. But the whole financial assistance should not be expected from the wife's family. As far as possible the wife also should have an self-governing income. Marrying a person who is very much above or below the economic and social status may cause several adjustment troubles for both the partners. In several love marriages, economic disparity may not be looked into which may result in marital troubles later on. It is very hard for love to flourish in a 'poor' circumstance.

Education and Intelligence

There should not be much disparity in these factors. In Indian villages, mainly women are still illiterate, men do not consider literacy as a pre-requisite for marriage. This is very sad state of affairs. Women should be educated to be intelligent wives and efficient home managers. An educated mother is the best insurance for the education of her children. As in the case of physical structure, men in India prefer partners with equal or a little lower stage of education. If the wife is more educated, the husband may develop an inferiority intricate. Concerning intelligence, both the partners should be intelligent enough to manage a home. Intellectual companionship flanked by the partners is a very significant factor which is not given adequate emphasis in Indian marriages.

It is not advisable to marry persons who are mentally retarded. The intelligence of the parents will be inherited through children to a very large extent. So intelligence is an significant factor in mate selection.

Attitudes and Values – Vital Orientation to Life

Areas where strong similarity is required are attitudes and values and the vital orientation to life. These contain goals, values, religious convictions, and general ideas in relation to the justice, honesty, truth etc. A simple tolerance of other's beliefs or conceptions is not enough to establish a life extensive partnership. The partners should fully agree with the thought of marriage, its meaning and purpose. It is very significant for them to have the same concept in relation to the fidelity, reciprocal roles in the family, sex, children, and in-laws. It is good to have a spontaneous agreement in fields such as entertainment, participation in cultural, social and political action, religious practices and community life.

Temperamental compatibility of the mates prevents troubles in marriages. Temperament means the physiological action stage and response patterns of the individual. The behaviour controlling glands of one individual may be dissimilar from that of another individual. Such biological differences help to explain several of the conflicts in marriage. For instance, why one spouse is always active and the other inactive. To avoid all these troubles, temperamental compatibility should be given due importance in mate selection. Compatibility of needs is another factor which should be looked into the choice of the mate.

There is proof that persons seek a mate who will compliment their strengths and deficiencies. Person with complimentary needs tend to marry. Several boys and girls are attracted to each other because each satisfies the other's need. The need so satisfied may be the need for love, affection, sympathy, understanding etc.

Now we have seen some of the significant qualities that are to be looked for in a mate. The selection of partner is not an easy procedure. It is an ongoing procedure which requires time and experience. It is better for the partners to have an opportunity to become well acquainted with each other. This described courtship. It provides the two persons the opportunity to test each other's feelings, ideas and orientations to life. This may help them to decide whether a life together will possible.

It also provides them occasions to reveal to each other the dissimilar sides of their temperaments. Inspite of the above advantages of courtship, it is usually not followed or accepted in Indian marriages. But it is a necessity in western countries. Anyway, it is good to have some acquaintance with the partner before marriage. One may not get a partner who can satisfy all the above criteria of mate selection. Some compromises have to be made in some characteristics. But once the selection is made, one should accept his/her partner totally. 'Accept toto' is a key word in marital success. Since it is not easy to change the behaviour of an adult the only way is to accept your partner with his/her merits and demerits and adjust accordingly.

There is no 'I' or 'You' in a marriage relationship, but only the 'We'. The husband and wife should become one and at the same time stay their own identities and individualities. The new Mathematics of marriage is 1+1 should be 'big one'; it should never be a -2- or a -11- where the partners proceed in two parallel rows which never meet.

MARRIAGE IN INDIA

CONCEPT OF MARRIAGE

Every individual has to play a number of roles in his or her life. Of the several roles one plays, two roles have a very great significance in life. One is the economic role and the other is the marital or family role. The former is prominent in life because one devotes quite a good part of one's carrier in performing it.

Consider one starts earning one's livelihood at the age of 20 to 24 years and continues to do so, up to the age of 58 to 62 years. That is, the economic career is spread in excess of to in relation to the four decades and that every day one devotes 8 to 10 hours to job/work. Thus one can well assume the period which one's economic role consumes in one's life. The marital role also involves in relation to the40 to 50 years of one's life. But of these two roles, the marital role is more significant than the economic role, because the latter involves secondary relations whereas, the former involves personal or primary relations. In order to understand this, more clearly, we have to see the variation flanked by primary and secondary relations.

Primary relations are essentially unlimited, particularistic, emotionally involved, altruistic and spontaneous. But, secondary relations are typically limited, standardized, unemotional, utilitarian and contractual. Again, primary relationship in marriage is dissimilar from primary relationship in other primary groups like friendship, neighborhood, village etc. It is in the sense that primary relationship in marriage is based on sexual relationship and this sexual relationship brings further intimacy and permanence in the relationship flanked by a man and a woman.

Primary relation in marriage has two significant functions: one of need gratification and other of social control. It gratifies biological (sex satisfaction), psychological (affection and belongingness) and economic needs (food, clothing and shelter) of the individuals and also acts as a primary source of morality and ethics. When one finds one's partner performing sure tasks for him/her, he/she considers it his/her moral obligation to care for the other or to listen to the other. One is, thus, no longer free to be immoral and irresponsible.

Another sociological way of conceptualizing family is how marriage involves performing new and varied roles such as husband, wife, father, mother, son, daughter, brother, sister and so on. Whether the persons involved are capable of performing those new roles or not, and how the inadequacy of performing these roles lead to family disorganization. What is significant in marriage is how the role enactment of one partner corresponds to the role expectations of the other.

According to Koos, a Sociologist, marriage is a dividing row flanked by the family of orientation and family of procreation in terms of the nature of roles one performs in the two families. The roles in the family of orientation vary in infancy, childhood and adolescence and carry no responsibilities and obligations. But the roles one performs in the family of procreation after marriage as a husband/wife, a father/mother, a wage-earner, a grand-father/grand-mother, a retired person etc. have dissimilar expectations and obligations.

Thus marriage is a miniature social system which necessity is kept in equilibrium if it is not to fall separately. Equilibrium requires adjustments, which in turn require provide and take or some sacrifice on the part of both husband and wife.

To uphold equilibrium it requires sure tasks to be performed through someone, for instance, of cooking, cleaning, wage earning, child-care and so on. Who performs which role is immaterial (though society has sure expectations from both husband and wife). What is important is that somebody should perform these roles for the stability of marriage.

VIEWS OF MARRIAGE

Several views have been expressed on this institution through anthropologist, sociologists and in religious texts. Marriage is an institution in which men and women are admitted to family life, to live in the intimate personal relationship, primarily for a purpose of begetting and rearing children. From the social point of view, marriage is an institution that serves to ensure propagation and socialization of children of a scrupulous society. From the individual point of view, it provides assistance in bearing and raising of children and controls for the receipt and extension of affection. Another view is that marriage is a socially legitimate sexual union, begun with a public announcement and undertaken with some thought of contract, which spells out reciprocal rights and obligations flanked by spouses and their future children.

Indian Views on Marriage

The above views represent western thinking in relation to the marriage. Now we are going to look at how Indian experts talk in relation to the marriage. Marriage has been measured a ceremonial gift of the bride through her father, or other appropriate relative to the bride groom in order that both may together fulfill their duties which are necessary for human subsistence.

These duties are "Dharma, Artha and Kama". Dharma is religious duty based on ethical values that is to do well and attain Moskha or salvation. Artha is the economic aspect of life and Kama is the physical or sexual duties to the partner and to produce children to perpetuate race. Marriage is not for sense of enjoyment, but to perpetuate the race. This is the Indian conception of marriage. It is a social duty towards the family and community. Indian religious texts say that marriage is a religious sacrament which is measured primarily a intricate of obligations, religious and moral on the one hand, and social and economic on the other hand.

The Hindu concept of marriage is that it is a sanskara (tradition) and a religious sacrament, not a contract. It is a holy union of the two souls and not simply of two bodies. It is an indissoluble bond which could be broken only through death. On the other hand Islam says that marriage is an institution ordained for the protection of the society and in order that human beings may guard themselves from foulness and unchastity.

Marriage is not a sacrament but civil contract, the objectives of which are the promotion of normal family life and the legalization of children. In the middle of the Christians marriage has been viewed as " a voluntary union for life of one man and one woman to the exclusion of others". This kind of marriage is monogamous.

Summarizing the above views, you might have learnt that marriage is a union of two persons of dissimilar sexes to lifelong reciprocal possession of their sexual qualities, which aims at the individual's biological, emotional, social and spiritual fulfillment and development and which cannot be achieved in separation.

Hindu Marriage as Sacrament

The Hindu view of marriage is that it is for fulfilling dharma and the need for pleasure (Kama). It is measured as a sacrament. There are many reasons for considering the Hindus marriage a sacrament.

- Dharma (fulfillment of religious duties) is the mainly significant aim of marriage
- Performance of religious ceremonies including sure rites like kanyadan, panigrahana, saptapadi etc. which are based on sure sacred formulae.
- The rites are performed before Agni (the mainly sacred god) through reciting mantras (passage) from Vedas (the mainly sacred scriptures) through a Brahmin.
- The union is measured in dissolvable and irrevocable and husband and wife are bound to each other not only until death, but even after death.
- Though a man can perform many sacraments throughout the course of his life, a woman can perform only one sacrament in her life i.e. marriage, hence it has great importance for her.
- Emphasis is on chastity of a woman and the faithfulness of a man.
- Marriage is measured a "social duty" towards the family and community and there are no ideas of individual interest and aspiration.

SHAPES OF MARRIAGE

Till now you have read in relation to the concept, meaning and dissimilar views on marriage. Now you are going to learn in relation to the different shapes of marriage. There have been dissimilar opinions concerning the original form of marriage. Some theorists say that primitive man lived in a state of group marriage.

The group marriage in which men in a group or tribe had indiscriminate access to all women of the group and children born out of this union were measured children of the general community. There are some other social scientists who believed that monogamy was its original form. Whatever may have been the original form of marriage, at present the mainly prevalent form is monogamy. But polygamy, polyandry levirate and surrogate shapes are also found.

Monogamy

Monogamy is the only form of marriage found in mainly societies; a man marries one woman, raises children within the wedlock and performs all rites with his mate. Monogamy has a extensive history of its own. Ancient Hindus regarded monogamy as the mainly ideal form of marriage.

Polygamy

Polygamy is marriage of one male with more than one female, or what may be described the "plurality of wives". A polygamous marriage may be unrestricted or restricted or conditional. In early Hindu society conditional polygamous marriage was practiced. According to the Dharmashastra, a man could marry again after ten years of his first marriage, if his wife is barren, or he could marry after thirteen or fourteen years if he had only daughters from his wife and wanted a son.

Manu said that, a man can marry another woman after eight years of his first marriage, if his wife is barren, after ten years if children produced through his wife do not remain alive; after eleven years if his wife produces only daughters, and immediately after first marriage if his wife is quarrelsome, rebellious, or harsh.

The Mahabharata says that a man who marries twice without any rational cause commits a sin for which there is no penance. Today polygamy has been legally prohibited. Besides the legal restrictions, people do not practice polygamy because:

- Maintaining higher living standards is not possible with more than one wife in the home,
- Plurality of wives augment tension in the family, and
- Women having economically self-governing status refuse to accept men's dominance in excess of them.

Polyandry

Polyandry is a marriage in which one woman marries more than one man. This was found in the middle of the Todas and Kotas of Nilgiris in South India. There are two kinds of polyandry which are prevalent in India. They are fraternal and non-fraternal. In the fraternal, the husbands are all brothers or perhaps from cousins from the father's side. In the non-fraternal, they are not related, as in the middle of the Nairs of Kerala.

In the 19th century the Nairs in the middle of the Hindus in Kerala practiced polyandry. But Westernmarck, a sociologist referring to these marriages, has said that polyandrous marriage unions of Nairs can hardly be described marriages because the male partners never lived with the woman and that the duties of fatherhood entirely were ignored. In 1896, the Malabar Marriage Act was passed which stabilized marriage norms in the middle of Nairs.

In the ancient literature the only instance of Draupadi's marriage with five Pandavas in Mahabharata period was justified through Yudhishtra on the foundation that similar marriages were performed through some of his ancestors and described it as "mother's command". Obeying the mother's command was a son's dharma. In the Mahabharata, referring to polyandry, it is said, "to have several wives is no dharma on the part of men, but to violate the duty owned to the first husband would be a great adharma in the case of woman".

Levirate

Levirate is a form of marriage under which a woman is taken as the wife of the late husband's younger brother or even throughout the life time of the elder brother and the younger brother exercises sexual rights in excess of the wife of the elder brother. The form prevails in the middle of the Ahirs of Haryana, some Jats and Gujars and some other castes of U.P.

Sorrorate

In sorrorate wives of a man are invariably the sisters. The origin of the word sorrorate lies in the Latin word 'Sorror' which stands for sister. In this form of marriage many sisters are simultaneously or potentially the spouses of the same man. It is usually observed in the middle of the Nagas, Gonds and Baigas of India who pay a high bride price. It has been observed that the death of the wife or her being barren is compensated through supplying a new spouse who is usually the younger sister of the deceased women.

MARRIAGE IN THE MIDDLE OF HINDUS, MUSLIMS, CHRISTIANS AND TRIBALS

In the Mahabharata four separate shapes of marriages are mentioned. They are: Brahma, Gandharava, Asura and Rakshasa. Gautama and Asvalayana, two of the ancient law-giver of the Hindus, refer to eight dissimilar shapes of marriage. They are Brahma, Daiva, Arsha, Prajapatya, Asura, Gandharva, Rakshasa and Paisacha. In the middle of them four were measured proper and desirable (dharmya) which had the approval of the father/family. The other four were regarded as undesirable (adharmya) who did not have the approval of the father. The proper marriages recognized through the "Smritis" were Brahma, Daiva, Arsha and Prajapatya while the four undesirable marriages were Asura, Gandharva, Rakshasa and Paisacha.

Brahma was the form of marriage in which a fully dressed girl – with proper decorations and ornaments – was given to a man of the same class through performance of the ceremony mentioned in the holy texts. The bridegroom used to be learned person of pure character and selected/approved through the bride's father. Daiva form of marriage was one in which the father of the girl gave her to a sacrificial priest as a part of his fee for officiating at the ceremony. The bride was properly dressed, decorated with jewellery which shaped part of his fees. In the Arsha form of marriages, instead of the dowry, the marriage was based on a system of barter in which the father of the bride was given a pair of cattle or two cows through the young man in swap for his daughter. The Prajapatya form of marriage is one in which the bridegroom is duly worshipped and married to the bride with due honors and blessing with these words, "go both of you and fulfill the duties of a householder".

In the Asura form of Marriage, bride-price is given through bridegroom to the bride's father. This is sort of economic contract. There is no limit of the amount given. The Gandharva form of marriage was the union of a willing girl with a man in solitude when both of them were in love. In this form of marriage, neither the consent of the parents nor the rites of dowry was essential. Only the will of the marrying parties was given importance. This marriage is believed to spring from desire and had sex satisfaction as its chief purpose.

The Rakshasa form of marriage was marriage through capture in which the girl was forcibly abducted through the man. If she cried for help and if her kinsmen came to her rescue, they were killed. The Paisacha form of marriage was marriage through seduction, stealing or fraud, where the girl was sexually violated while she was asleep, intoxicated or unconscious or when incapable of protecting herself.

Of these eight shapes of marriage Brahma is measured to be the best marriage, where a girl is married to a boy of merit in the same caste or in a caste of equal status. In the Mahabharata age, the two mainly prevalent shapes of marriage were Brahma and Gandharva. Gandharva marriage was declared to be proper for a Kshatriya. Several of the heroes of Mahabharata contracted this form of marriage; for instance, Arjun married Ulupi and Chitrangada, and Dushyanta married Shakuntala. Swayamvara: A variant of the Rakshasa form marriage was measured to be the norm for princely homes; for instance, Arjun won Draupadi and Nala won Damayanti.

Marriage within the caste was the prevalent order of things throughout the period of the Smritis and the Puranas. Marriage was rigidly regulated through the caste system and caste laws. As a matter of fact, a man of higher caste could marry a woman of lower caste which is described anuloma marriage.

But a woman of high caste marrying a man of lower caste which is described pratiloma marriage was not allowed. Another custom relating to marriage was the law of consanguinity – close relationship through blood in excess of generations. There are sapinda, gotra and pravara degrees of relationship. They are planned to prevent marriage in the middle of sure kins and gotras. The present custom of Hindus is that marriage within the same gotra is to be avoided.

Endogamy

Endogamy is a social custom that requires a person to select a spouse from within sure groups. These endogamous groups specifically refer to Varna, caste and sub-caste. Thus, a boy from a scrupulous caste marries a girl from the same caste. Caste endogamy was functional in early society because:

- It made marital adjustments easier,
- It preserved the occupational secrets of the caste,
- It maintained the solidarity of the caste, and
- It checked the decrease in the membership or strength of the castes.

The mainly typical endogamous rules are enforced through tribe race, religion, and social class. Hindu couples planning to marry are obligated to belong to the same caste and sub- caste. Sub-caste refers to a further subdivision of castes into endogamous categories which, for all practical purposes, are themselves self-governing castes.

The negative effects of caste endogamy are that, it creates:

- Inter-caste tensions which adversely affect the political unit of the country,
- The problem of marital adjustment, because the field of selection remnants limited, and
- Troubles of child marriage and dowry system.

Exogamy

The term 'exogamy is essentially sheltered through the incest taboo which is prohibited universally. Likewise, universally banned is the marriage flanked by brother and sister. An motivating exception to the latter prohibition occurred in the royal families of ancient Egypt, Hawaii and the Incas of Peru. The explanation for this is the determination to uphold intact a royal row of descent in societies which had not urbanized a system of inters marriages with foreign royalty. Close blood relationship is typically a bar on marriage for all people in all times and places. Regularly first cousins may not marry. Alliance of affinity, as well as blood relationship, has been incorporated within exogamous rules. In the middle of some people, marriage flanked by persons belonging to the same village or other territorial group is banned, or at least discouraged. Exogamy is a social custom which forbids selection of a spouse from sure groups. There are two kinds of exogamy practiced through Hindus. They are Gotra exogamy and Sapinda exogamy. They are planned to prevent marriage in the middle of sure kins and gotras.

- Gotra Exogamy: Gotra is a group whose members are believed to have descended from a common mythical ancestor. Initially there were only eight Gotras, but slowly their number increased to thousands. The Gotra exogamy prohibits marriage flanked by members of the same gotra.
- Sapinda Exogamy: Sapinda means one who carries the particles of the same body. Sapinda relationships arise from being linked through having particles of the same ancestor. Marriage flanked by such persons is prohibited. Since there is no limit to persons related through blood, some limit is prescribed for avoiding persons for

marriage related to each other within sure generations on the father's and mother's side. In practice and according to law, five generations from father's side and three generations from mother's side are avoided. Though, breach of sapinda exogamy was never penalized, though breach of gotra exogamy was measured a heinous practice.

Cousin Marriage

There are four kinds of cousins:

- Chachera (father's brother's son/daughter),
- Mamera (mother's brother's son/daughter),
- Phuphera (father's sister's son/daughter) and,
- Mausera (mother's sister's son/ daughter).

Of these, chachera and mausera cousin (where the two sibling parents of the child belong to the same sex) are described parallel cousins and mamera and phuphera cousins (where the two sibling parents of the child are of opposite sex) are described cross cousins. Of these two shapes of cousins, cross cousin marriage was practiced in ancient Hindu society. Even now cross cousin marriages are practiced in the middle of Hindus and Muslims.

The main arguments for and against cousin marriages are biological, social, psychological and cultural. The arguments against cousin marriages are:

- It will lead to biological degeneration of family because parental defects will be transmitted to their children,
- It will create secret relations flanked by primary relations in the family and thereby lead to immorality; and
- It will be against our religious norms.

Arguments in favor of cousin marriage are:

- One's property will remain in one's own family;
- It will create stronger bonds of love flanked by brother and sister, and
- With the breakdown of joint family cousins no longer live together in the same home.

Hypergamy and Hypogamy

As a matter of fact, a man of higher caste could marry a woman of lower caste which is described anuloma marriage (hypogamy). But a woman of high caste marrying a man of lower caste which is described pratiloma (hypogamy) marriage was not allowed.

Inter-caste Marriage

It is a marriage flanked by man and woman belonging to dissimilar castes. Inter-caste marriage in India is usually understood to mean not only marriage flanked by sub-castes of a major caste group (as flanked by Brahmin sub-castes), but also marriage flanked by two major castes (as for e.g. Brahmins and Vaishyas).

Inter-caste marriages which are increasingly common in India, especially in urban and industrial areas, are not flanked by high and low caste people but flanked by the members of several sub-castes within a large caste group (as flanked by Brahmin sub-caste). At present, several of the modern minded and liberal people in India believe that inter-caste marriages should be accepted (legally such marriages are permitted) through the people/ society. Also they believe that inter-caste marriages would help breakdown the traditional caste system. It will also help wipe out caste distinction and untouchability.

Formerly marriage outside one's caste was not to be even thought of. Today several men and women are prepared to break through the bonds of caste if mutual love or attraction demands it. Though, studies mannered on inter-caste marriages show three characteristics.

- When a person outside the caste is wealthy and had a social prestige, there has been a general approval.
- Persons with higher education and who are older marry outside the caste.
- Even today, the large number of marriages is within the caste and these are marriages arranged through parents. The only important change is with respect to the restriction against sub-castes which has now been practically eliminated at least in the urban areas in the middle of the educated persons.

Arranged marriages are the marriages arranged through the parents of the bride and the bridegroom, considering all the norms and customs of the society and religion.

Inter-religious Marriage

Inter-religious marriage in India is basically understood to mean marriage flanked by persons belonging to dissimilar religions.

Power of Legislations on Hindu Marriage

In excess of a period of time several beliefs, values and ideals related to Hindu marriage have lost their original meaning and purpose. People started questioning sure evil practices like child marriage, sati system, restrictions of widow remarriage etc. consequently throughout the British rule and even after independence several legislations were enacted.

Some of them were; The prevention of Sati Act, 1829, The Hindu Widow Remarriage Act, 1856, The Civil Marriage Act, 1872, The Child Marriage Restraint Act, 1929 and its amendment in 1978, The Hindu Marriage Act, 1955 and The Dowry Prohibition Act, 1961 and its amendment in 1986. The above mentioned legislations have led to important changes in the Hindu marriage system. Some of the significant impacts of legislations on the institution of Hindu marriage are:

- Divorce is now socially and legally permissible. Thus marital relation has ceased to remain unbreakable.
- Provision for widow remarriage and divorce has affected the ideal of 'pativrata'.
- Marriage is no more a religious duty rather it is performed for lifelong companionship.

Marriage in the middle of Muslims

Marriage in the middle of Muslim is universal and obligatory. Since the Muslim community discourages celibacy, marriage has to be performed. It is true that marriage in the middle of Muslim is a civil contract as it is meant for procreation of children and legalizing sexual intercourse; it is a religious duty also. It is measured an 'ibadat'. Feature characteristics of Muslim marriage are; acceptance of the proposal of marriage through the bride; capability of the bridegroom to enter into a marriage contract; preference system i.e. parallel cousins (father's brother's daughters) and cross cousins (mother's brother's daughters) are given preference; and marriage is valid only if it is free from legal complications.

It may be noted that man and women did not enjoy equal rights with regard to the provision of divorce. Women have always been at the getting end. Though, industrialization, urbanization, modernization and spread of modern education have drastically changed the perspective. Spread of small family norm and lesser incidence of divorce are some of the impacts of social change on the institution of Muslim marriage.

Marriage in the middle of Christians

The Christian community has two major denominations: Catholics and Protestants. The Catholic owes allegiance to the Pope. The Pope is the supreme power in the Catholic Church. All the teachings of the Catholic Church have the approval of the Pope. The Protestants have many denominations or groups. Hierarchical approach is limited within each denomination. As per the teaching of the Catholic Church marriage is a sacrament. There is no provision for divorce. Though a marriage can be declared null and void if one of the spouses is already married and the partner from the first marriage is still alive. A marriage can also be declared null and void in case if the spouse is of unsound mind, impotent etc. at the time of marriage. But the procedure to get a marriage declared null and void from Church is very tedious as the clearance has to come from the Vatican.

In the middle of the Catholics mixed marriages are permitted (with a person from any other religion). In the middle of the Protestants divorce is permitted.

Marriages are usually performed within the church in the attendance of relatives and friends. Married couples are free to register their marriage for legal purpose. Protestant couples usually seek divorce from a court of law. In the middle of the

Protestants, remarriage is also permitted after divorce. According to the teaching of the Church free consent from both the parties is necessity. Consent necessity is an act of the will of each of the contracting parties, free of coercion or grave external fear.

Tribal Marriage

Family comes into being only through the establishment of culturally controlled and sanctioned marital relations. Marriage, so, is universal. We find several shapes of prescriptions and proscriptions concerning marriage in the middle of tribes, also.

Though, uniqueness of tribal marriage lies in the ways through which mates are acquired.There are: probationary marriage in the middle of Kuki; marriage through capture in the middle of Nagas, Ho, Kharia and Birhor; marriage through trial which is recognition of personal courage and bravery in the middle of Bhil; marriage through what has been described purchase or bride price prevalent all in excess of tribal India: marriage through service—a solution to the problem of high bride price in the middle of Gonds and Baigas; marriage through exchanging women of two households for avoiding the payment of high bride price prevalent all in excess of India except Khasi Tribe; marriage through mutual consent and elopement in the middle of those tribes who have youth dormitories, and marriage through intrusion in the middle of Birhor and Ho.

SOCIETY, CULTURE, RELIGION AND FAMILY VALUES

FAMILY LIFE: CHANGING PATTERNS

Definition of Family

A family is a community of persons related to one another, living together in an environment of understanding and acceptance. A family is a place where one can freely express one's feelings, emotions and needs without being threatened or ashamed of them. Family is a place where one can feel security, wholeness and a sense of being wanted.

Family can also be defined as "Those related persons who live together within a household, usually with common eating habits or one kitchen." It is experienced and proved that children grow best in an atmosphere of security and affection and that material attributes can never be substituted for true love.

Kinds of Family

Development in society in the recent years have focused their attention on man's self-fulfillment through individualism, materialism and consumerist values isolating man from his need to belong to one another as persons. Authentic fulfillment and happiness is experienced not in the acquisition of the external, material wealth and possessions, in accessible from one's relations to the rest of humanity, but in his intimate and important experiences of other people starting with the family.

In fact every human being is bound together in family and in the earliest experience of interacting with the father, mother and other important people, one learns to find meaning and identity for the rest of one's life. The traditional pattern of family living in India was that of joint family, in which members were bound together through ties of common ancestry and common property. Now in India we find three kinds of family structures.

- The large Joint Family
- The Nuclear Family
- The Extended Family

A joint family is one that has a greater generation depth than the nuclear family, in which members are related to one another through property, income, mutual rights and obligations. The care and maintenance of dependents is a moral obligation. Members of the family are closely knit together and share the troubles and joys of social living, having strong feelings of mutual obligations throughout crises and regard self-interest as being identical with family welfare. The joint family provides an "umbrella" of support, which covers financial loss, decease of a spouse and social security.

Children brought up in a joint family are more secure, affectionate and have a well-urbanized personality. Joint family can be found more regularly in the middle of Hindus than amongst any other community; in the middle of agriculturists rather than traders, clerical and professional workers. Joint families can be found in the lowest income groups too. Men grow up with the knowledge of people approximately, confidence and social in family business matters. Children grow up with the experience of life cycles: birth, maturation, marriage and death. There is no difficulty of boy-girl relationship, no problem of single woman, no problem of single or neglected child and marriage are arranged through parents. The positive characteristics of joint family system are based on the advantages of the members such as:

- Family survival
- Care in old age
- Augment in family income
- More share in the property
- To follow the pattern in the community
- To make the home happier

Nuclear Family

Nuclear family is one in which the parents and their unmarried sons and daughters live together.

Extended Family

The extended family is a later development of the joint family system with a transformed image. It mediates the nuclear family and the large joint family. In this kind the nuclear family is extended with sons marrying, bringing up children and remaining within the original family of the parents.

Traditional in Family Patterns

We live in a fast changing world. We have reached a time in history when sitting at home, we are able to know what is happening approximately the world. We are able to move approximately the world in a shorter time than before. We are living with all the modern technologies that dictate easy life, pleasure, comfort and give automatic answers for mainly of our troubles. Several are doing their own business sitting at home. The science and technologies have progressed to such an extent that man only has to sit in homes press a button in order to send a written message crossways the world and receive the reply. Any information one needs on any topic can be obtained through our family computer. In excess of the years, the Indian Family has gone through several transitions.

The superior families of six and eight children have been replaced through families of one or two children who learn constantly to compete with each other in excess of toys and personal possessions from their infancy. In place of joint families where adults, children and the elderly interacted with one another in a secure atmosphere, now we have strangers and lonely individuals living in separate worlds.

The elderly are conveniently put into old-age homes and children are kept waiting at the school gates or in front of family T.V. till their parents return from the offices. Children who spend extensive hours in closed homes or in the company of servants or other school children develop a sense of rejection, depression and separation. All that they are able to interact with for extensive hours after school are toys, story books, T.V. computer games and cartoons that contribute to the formation of a mechanical, dehumanizing, individualistic, and narrow-minded attitudes in children.

Significance of Transitions in Families

The joint family cannot be dismissed as outdated. Several Hindus as well as Muslim still families follow that life-style. Rank and wealth are not the circumstances for this system but blood relations and social value are. They are mostly found in non-urban, non-industrialized settings. It is motivating to note that even when the members of an extended family do not live in the same household, they still share a common budget and follow the same family leader. Besides, even in the middle of the members, who live separately, we observe that all belong to a joint family system and believe in this value.

There are troubles linked with all extended families, as the joint family is based on the relations in the middle of the adult males rather than on the conjugal bonds flanked by spouses.

The spouses do not feel the allegiance toward the large units as the husbands feel and there are quarrels, competitions and dissatisfaction in the middle of women and children. In recent years majority of Indians prefer to live in nuclear families. The role of religion is to provide a perspective to human life, hence, to family life and through it to society.

CHARACTERISTICS OF VALUE FORMATION IN FAMILY

Definition of Values

Values are those ideas which direct our thinking and living and provide meaning to our subsistence. The values we hold dear are expressed in our actions through which our lives receive a sure amount of identity. It is the value one possesses which gives him/her an image through which others judge or measure him/her. In order to make values meaningful, they have to be practical and directive.

Classification of Values

Values can be classified into three groups—personal, social and neutral. This classification is based on the traditionally held on customs and beliefs inherited through the person which make a society proud and protective of the culture which is inherited. Values can also be classified as material values and higher values. Material things have their own values: food, physical health, and clothes are such visible values through which one exhibits one's identity and are essential for survival in a society.

Higher values are seen only through way of behaviour. Since food is a necessity for human life, working for daily food is a value, but to share what one has with other needy people is a higher value. Even though joy is a value, all things which provide joy need not be values, though there can be a desire to possess all things. Higher values are eternal values: they can be understood as human or divine.

Whenever, there is a doubt; a need for the expression of values becomes apparent. Life is made up of small and big decisions. Where families are authentic in loving one another and caring for each other, obeying God and working for Him as the source and centre of the universe, progress, peace, harmony and human development will be the outcome.

Socialization and Value Formation

Social, religious, and cultural values affect family life to a great extent. In fact no family can exist independently of these factors. One of the mainly important stages of one's life is socialization. Socialization is a procedure through which the child absorbs, assimilates and internalizes socio-cultural and religious practices from the important persons in his/her life e.g. parents, teachers, siblings. Thus, a child acquires attitudes and values, some of which pertain to moral standards and others relating to people love and hate, superior and inferior, etc.

These attitudes and values exist in children in dissimilar ways, at dissimilar stages according to the atmosphere, psychological stages and physical circumstances.Socialization is seen through sociologist as a procedure toward cultural conditioning. For an individual infant to survive, it necessity be socialized. There is a structural relationship socialization. As result of being socialized through parents, one becomes an agent of socialization for one's children and grand children. These obligations become fixed and specified for generations. It is easy to understand that such a conditioning attached to a joint family system can be so strong in the Indian society as it has lived for in excess of 2000 years with its culture, beliefs and socialization procedure provide a meaningful values to family life that are inseparable from society.

Thus we see that the individuals and the families are motivated to serve the society's interests and stay up the system. In this sense the socialization procedure is more or less a natural procedure to make the individual conform to the social norms and cultural pattern of a society. There is a tendency in several individuals, to deviate from this kind of forced pattern, since all human beings have an inmate longing for freedom and self-expression, for spontaneity and personal identity. This explains the cause for teenagers and young adults rising up with antagonism and rebelling against parents and those in power.

Value Formation in Parent Child Interaction

The mother-child ties are emotionally intimate which facilitates the procedure of socialization. The span of human life is extensive and hence socialization continues. The pattern of dominance through father or mother gives further power to what is learned and fixes attitudes and values in the minds of children. Already learned experience of children through families is made forceful and commanding through the way the society lives and interacts. At this stage, the experience of the father and mother and the strength of the parent child relationship will determine the values of selfhood (identity, security and self image of the child), which form the core of a child's personality.

Children of pre-school age are already exposed to a competitive world. Parents, out of enthusiasm for proving their identity and getting approval for their status, pressurize their children to achieve excellence in studies, sports, music and other fields and push them beyond their capability and speed up their performance. But slowly this creates stressful feelings and inferiority complexes in children and they become victims to physical and verbal humiliation in school and at home.

Women's Role Affecting Family Values

We cannot satisfactorily talk in relation to the formation of family values, through passing the nature and role of women in both society and family. The home is the pivot of society and the woman is its centre. The home – the family – is the first institution which imparts self identity to an individual and where he/she formulates a self image. The respect and esteem the mother is given in the family greatly determines the values with which the family looks at others in society.

"A women's position in the family is lowered if she gives birth to female children, and she is looked down upon through others. On the other hand producing sons is measured to be a great attainment". Such families cannot provide a positive self-concept to female children who in turn look at women negatively and pass on negative values to the coming generations. "The cultural conditioning has gone so deep into the psychology of people of India that even in the middle of the educated and so described enlightened and well-to does urban-dwelling people, the birth of son is much more desired".

Though some change is apparent in cities, mainly of the village women still live under the subjugation of men and traditional social customs. It is clear from what has been said that one of the significant characteristics of family values depends on the position that women occupy in home and in society. The women's self-image, the opportunities and congenial environment for utilizing the rights and privileges due to her has a great deal to do with the kind of families and the excellence of values imparted to the family members. Hence women's empowerment is necessary for the formation of new values in Indian families.

FAMILY VALUES AND POWER OF SOCIO-CULTURAL AND RELIGIOUS DIMENSIONS

Values in Family and Society

The family is made up of individuals but it is also a part of the superior social network. Thus, individuals are initiators and promoters of culture. Individuals and society are mediated through families. It is in and through families that people learn and pass on values. But all members of a society are under the constant supervision of parents in childhood, of friends in teenage, and of the public in adulthood. Family is where children are trained; the adults feel free to praise, criticize, suggest and order so that the children learn family preferences. They are taught what is right and wrong, what to tell others and not, how to behave inside and outside the home, who necessity be their friends etc. All these power and communicate to the children directly and indirectly the attitudes, values and conduct which remain with them for the rest of their life. What they learn is part of that culture and customs of the society which are valued through the adults.

Family as a social institution is the vital structure of the society. Hence the flow of values flanked by the family and society is very lucid and they interact very closely, intrinsically and inter-dependently before the effect is apparent. It is not wrong to say that it is through the family that the values and attitudes are absorbed through the child in his/her early age, cultural transmission of scrupulous society is kept up and the major agent for all these is the woman. The significance of the family is the mediating function of family in the superior society, as it links the individual to the superior social structure. As part of the society, families like to stay their identity connected to it, hence the family values remain mostly undifferentiated from the values of the society.

Socially, man's relationship with other human beings is diminishing. His engagements are more with the machine and technically produced means of communication. Communication, which is the medium of relationship and social connection, is no more concerned with human relationships, but media and media-related learning and knowledge. Media also has helped to replace human interactions with group interactions.

Studies have shown that in societies where human life and peaceful living are valued, there is little competition and there seems to be hardly any use of one another. Work is done essentially in co-operation and there is no economic rivalry. Women are respected and incorporated in the decision making processes. We can also see that the opposite is true in societies where wealth and success are valued in excess of human life. When money, position and power are valued, individuals grow up with aggression, violence, competition and manipulation. Families become unable to foster human values that can sustain and uphold members, as they are subjected to the negative powers of peers, employees, neighbors, and organisations in the society. According to Mahatma Gandhi, truth and love are the mainly admirable and cherished values of youth.

Values in Family and Culture

The Indian family is no more unicultural following customs, practices and beliefs of definite pattern but a plurality of practices and patterns. The elders used to be the central persons whom the youngsters looked up to, learned from and were agents of imparting the traditions and values of family, life, parent-child relationship and discipline. The uniqueness of the Indian culture is being engulfed through the global culture. Though, the villages still hold on to the traditional values of hospitality, simplicity, submissiveness, and belief in God.

They have been excluded from the modernism through deprivation of economic and technological development.

Man's uniqueness consists in developing language and preserving meaningful, creative behaviour and in communicating the internal circumstances and feelings through external expressions. The cultural implications of values are tied up with practices, symbols, religious rituals and the customs of a society. Also, traditions and languages of the dissimilar regions or nationalities are expressions of values of the ethnic group to which one belongs.

Though India is recognized for its unique culture, the dissimilar religions have specific cultures. Culture expresses itself in one's life-style, food habits, dress and symbolic expressions. Indian culture is unique in comparison to the West, but each culture has its won richness and inherent in it are specific messages, interpretations and expressions of values the people of that society hold dear. The diversity of perception, experience expressed through dissimilar and creative ways become integrated in the life of individuals through learning, understanding and accepting in the given cultural contexts. The family is the place where the atmosphere of unity, integration, harmony and respect of dissimilar beliefs and expressions become a necessarily parts of unified consciousness.

India is a land of several striking contrasts, and a great cultural complexity, social diversity and regional variation can be found in the middle of the people. And in the diversity of the Indian scene it is very hard to make any generalized judgments concerning the values in family life. But it is the unity running through diversity which reveals the positive dimensions of social and cultural values in family life. Older attitudes and prejudices still persist and social opinion is taking its time in changing itself. As values are integrated with culture, religion as well as socially determined attitudes, behaviour and customs, an evaluative and critically questioning methodology is required for the promotion of family values. In course of social change people adopt new ways of living. Old customs and practices fail to make meaning to younger generations. Hence, the need for introducing meaningful values becomes a necessity. For instance, the role of women was measured to be within the family, taking care of the husband and children, but now as women are working on equal terms with men, outside home the attitude toward women cannot remain the same. This perception affects family values.

Values in Family and Religion

The world is not the ultimate reality. This is the teaching of all Religions. The Bible, Gita, Koran and other scriptures affirm and acknowledge the one and only powerful and omnipotent God who is invisible to the external eyes and to the physical world in which we live. All religions emphasize the need for forming a mature conscience in people. Religion provides a code of norms which will guide and enable persons to have an objective understanding of God. Religion necessity help people to grow in a balanced, harmonious, altruistic attitude which will result in accepting all people as one's relations, regardless of their caste, creed, religion, or customs/ language.

It is hard to differentiate the socio-cultural and religious values as they affect family life. Religion in its purest form deals with one's faith in God and shows how he/she necessity live in this world. Hence religion frames 'law' or teaches one's duty to God and towards others, which we call Dharma. All religious founders taught how to live our lives in worship to God and our duty towards our neighbor. Hence Buddha taught compassion and less desire for worldly pleasure. Jesus taught of love for neighboure to the extent of giving up one's life for the other in service. Hinduism talks of Nishkamakarma; doing one's duty toward others and not expecting the results of the actions. Though, the truth is that it is the religious beliefs of a society that sweeps under all other values in family and society.

Family Values and Attitude toward Sex

According to the study and analysis of number of psychologists and educationists, the attitudes and values (either positive or negative) which the parents have, become the mainly powerful instrument in the hands of the children for later years. The self-image, as well as the ability to interact with other in children from childhood to later years is affected through the early experience of sex-related values taught in the family. According to a classic concept a little boy at the age of five or six chooses his mother as the first substance of his sexual desires. The same thing is true in relation to the girls concerning their father. The upbringing of children with a positive, balanced, relationship with family members and outsiders at this age is essential for the growth of proper values concerning sexuality in children. The understanding of sex as part of the body for a special purpose in the plan of God for the world has to be taught to children slowly as they mature in life.

Respect for all persons, regardless of sex variation is a higher value which very few people possess. At the same time a family that does not respect persons and considers sex as an substance of pleasure and a means of play can cause negative understanding of sex in children who may grow up to be exploiters of women in society.

Values in Family and Role of Media

As we have already seen earlier, with the breakdown of the joint family and the advent of industrialization, the family atmosphere has changed drastically. The family has become small in size, the elderly and children are left alone at home and the parents are burdened with in excess of time jobs. The rise of materialism and consumerism has created unnecessary anxieties of separation and loneliness in families and in the middle of families. The advertisements set the norms for values concerning food, dress and friends. The concepts of values are created in children through what they see and hear in the media. T.V., internet, computer and other technological devices have become source of entertainment for both children and adults. In place of listening to the stories of parents and grand parents while going to sleep, children seek music, serials and cartoons before going to bed. Role models for the present generation are film stars who constantly appear on televisions. Balance in the media against the values of families has gone beyond the control of parents.

On the positive side, children are becoming aware of the wider dimension of human life and interactions. Young children are rising rapidly in their intellectual curiosity and general knowledge of the world approximately them. Teenagers, youth and adults themselves are becoming more aware of their rights. Environmental Preservation, Animal Protection and Human Rights are becoming issues of priority for the young people of today. The demands of children and adults in families to upkeep with the media-world have become an obsession affecting the dignity and self-worth of family members.

FAMILY AS THE AGENT OF A NEW SOCIETY

Family in Procedure of Value Formation

The truth that family is the foundation for a stable and strong society is becoming more and widespread throughout the developing countries. The topic is complicated one, as family, society, culture and religion are inseparable factors and each one has something to do with the shaping of values in the individual and families. The family is a 'mini school' and it is there the children learn to relate with others. According to physiologists the child perceives and experiences its parents in the first year of life. The child's personality and attitude to a great extent are shaped through the age of four. The important persons of the child's life in this stage are the mother, father, brothers, uncles and aunties. This is the period in which parents can play a great foundational role in families.

The future of humanity passes through way of family. Philosophers and social analysts have noted that society is a structure made up of families, and that peculiarities of given society can be described through outlining its family relations. The earliest moral and ethical writing concentrated much on family due to its importance as the base of the society. True happiness and progress of a society depends on the positive attitudes and behaviour of the members of the family. When these attitudes and relatedness extends from family to society, the society mirrors the family. When the power is vice-versa, the family becomes the mirror of the society. In either situation, both family and society power each other and one of them will always be dominant.

Family necessity become the Vital Unit of Change

Families as the vital and mainly significant unit of social institution necessity are given special attention through any educational field that deals with society and progress. Being the fundamental constituting unit of the society, the family concerns itself with the well being and growth of persons. A happy family is where the parents and children together experience the joy of belonging, and the feeling of being cared for the deepest needs. Food, clothing and shelter are the vital needs of all human beings, but the happiness of a family depends on higher needs such as being loved, trusted, understood, for given, recognized for one's talents, gifts and the ability to reach out to those less fortunate.

The fulfillment of family life is in bringing forth persons strong in body, mind and spirit, with emotional maturity and altruistic outlooks as citizens of the nation. Persons necessity be socialized and educated for bringing in relation to thee social change. Hence, schools have a vital role in the attitudinal change and inculcation of values in children. This has to be done in co-operation with families. Parent-Teacher Associations and Management Committees are such means through which schools and families can take co-operative responsibility for education of children in the true spirit of formation.

Personality Traits and Vital Life Orientations

- Vital trust vs. vital mistrust;
- Autonomy vs. shame and doubt;
- Initiative vs. guilt;
- Industry vs. inferiority;
- Identity vs. role confusion,
- Intimacy vs. separation;
- Generativity vs. stagnation; and
- Ego integrity vs. despair.

Men and woman are made in the image of God so that they can think, question and differentiate flanked by good and bad, design their own destiny and reach the ultimate goal.

Parents' Focus on Value Clarification

In order to function as moral human beings in society, children need to grow in an atmosphere of clearly defined values which are practiced through adults. Children need to see/hear parents doing and talking what they believe in. Mostly, children learn from adults through imitation when they are young, and through reasoning when they grow up. They look for impartiality in behaviour and preaching in practice. Hence there is a need for clarification of values in the day-to-day life of the families. Initiated through parents, children seek guidance and answers to question in order to understand and accept values in the light of their purpose in life. So, every family's first priority in training and educating children necessity begin with the question of personal identity.

Who am I? Where have I come from? Where am I going? This vital foundational principles of life necessity become a stepping stone for the positive socialization of children. Religious and moral values of God as the Creator-Father and the Universe as His Creation – Home should occupy significant place in their value domain.

A relationship with self, with God and with others will pave the way through to the child's mind. Children then will learn to relate to others as extension of themselves and develop social consciousness. The understanding of personal, social and spiritual relationships will grow in children through their experience in the family with the interaction with one another, day in and day out.

Family Values and Positive Strokes

Psychologists have come to the conclusion that positive thinking is the secret of happiness. In order to grow up in positive thinking, one necessity is exposed to positive strokes. The families where parents accept one another and encourage what they are and their contributions, health and happiness abound. From childhood on, children need to feel secure in themselves from the recognition and understanding they get from their important people. Positive strokes can be effective in families for the proper self-image and upbringing of children.

Case studies show that the mainly unhappy and violent people have been those whose childhood was spent with adults who were strict, constantly nagging, criticizing, and condemning.

Allowing children to grow up with clearly defined freedom, respect, love, encouragement and appreciation will result in making new generations of good citizens for the nation. The excellence of change thus brought in relation to thin persons through families will prepare the groundwork for causing a gradual change in the society. So, the parents necessity concentrate on giving positive strokes to the children in order to direct them toward a new value system making families agents of social change.

Marriage is a point of departure in which boys and girls, brothers, and sisters realize themselves as persons with responsibilities in life, from individual ties with other persons and appear mature and able to make decisions for life.

Recommendations

In order to form families with lasting value that can power society and change it for the better, the following values necessity be inculcated and transmitted to children both through parents and social institutions.

- Family life necessity is based on truth, faith, and justice.
- A culture of peace, brotherhood, and tolerance necessity is taught in families.
- Respect for life and respect for dissimilar views regardless of religions, ethnicity and regions necessity be fostered in families.
- A sense of responsibility for the welfare of others; family necessity be recognized through society and educational systems as a primary unit of society.
- Adult's necessity become examples of inter-faith dialogue, critical analysts of social issues and promoters of Nation Building.
- True patriotism necessity is affected in children through parent's efforts and social institutions.
- Education at home and in an institutional set up necessity directs the yearning of human communion and self-realization.
- Young people necessity is recognized for their energy, enthusiasm and new knowledge and the elderly necessity are respected for their wisdom and practical knowledge.
- Correct judgment and common sense necessity be cultivated in children right from childhood.
- Opportunities for making personal decisions and evaluating the choices be made and accepted out.
- Children necessity is brought up with consideration for their aesthetic, moral, intellectual, and emotional development.
- Boys and girls necessity be brought up at home with equal participation in the household duties, given equal treatment and respect.
- A sense of compassion for the elderly and those less fortunate due to cosmic, social, or economic situations necessity is part of children's curriculum.
- Sacredness of sex, beauty of rising up and happiness of family life necessity become experiences for children in families.
- Lastly, there necessity is the conviction that every parent in this world has a privileged duty of being an agent of re-creating the society and transforming the world.

MARITAL LIFE AND ROLE EXPECTATIONS

ROLE EXPECTATIONS IN MARITAL LIFE

Shakespeare wrote that the world is a stage upon which men and women are acting out the drama of life. The same thing can be said of marriages. There are several cultural and social expectations in relation to the appropriate behaviour for males and females. The goals, purposes and functions in marriage can be achieved only when each family member plays his or her scrupulous role.

Marital Roles

What is meant through marital roles? Marital roles are the behavioral expectations of husbands and wives in a scrupulous society at any given time. Roles are cultural creations, rather than biological imperatives. Roles give the facility for the smooth running of society through a division of labour for men and women.

Changes in Gender Roles

Are there any hard and fast rules for assigning sex kind roles? Can we change the gender roles? Of course there are no definite rules to assign roles to men and women. Cultural variations and the purposes of spouses may bring in relation to the changes in marital roles. But they cannot violate nature's goals for marriage. So we can find that some marital roles are unchangeable.

Disparity in Role Conceptions

In the past there were clearly defined roles for man and woman. Today we all are living in a rapidly changing society. In the modern society, there are no definite patterns of behaviour or roles for men and women. There is wide disparity in role conceptions. This changing nature of gender roles creates troubles for all kinds of couples as they settle down to live together.

Factors Affecting Changes in Roles

The industrial revolution, emancipation of women, urbanization, employment of women, preoccupation of men with career etc. are some factors which have left their imprint upon family roles.

Role of Man

Role patterns for each sex are based on masculinity and feminity. Men and women are born with sure vital features. Each culture has its own expectations of the roles of males and females.

- Unique male roles: What is the key role of man? On what foundation do we assign this role to the male? In mainly cultures strength and courage are still measured to be the qualities of men. They are also self-governing, tough and can control their emotions in better ways than women. Physical strength and social dominance are more functional for the male roles. So man is usually expected to be the provider and protector of the family. He is the master of the family. According to the traditional role expectations, he is the bread earner and the head of the family. He is expected to acquire a job to support his wife and children. He has to show determination in the attainment of hard goals.

- Leadership role: The man is also entrusted with the role of leader and supervisor of all family endeavors. For the child, his/her first heroes will be his/her own parents, particularly the father who holds the position of power in the family.
- Role as a husband: As a husband he has the role of sex partner, companion, confidant, decision maker and accountant. He necessity train himself to be a better observer so that he can be of great help to his wife. His necessity notices his wife and praises her performance and ability. He should also provide emotional support to her.
- Role as a father: The father holds a dominant position in the family. In our culture, he is the chief power in the home. Children need him for their all round development. They learn several good qualities from him. Some of the qualities are sense of justice and fairplay, steadfastness, inspiration to be useful etc. Hence he should be firm without being despotic, decisive without being dictatorial and gentle but not weak.

Role of Woman

Females, in mainly cultures, have been conditioned to carry out the reproductive function.

- Traditional roles: The woman is biologically, psychologically, and emotionally prepared for motherhood. She is trained to carry out the roles of birth, nurture, protection, gratification and giving comfort to children and men. In the life of woman, these functions are given priority in excess of all other engagements. According to the traditional role expectations, she is oriented towards rearing capable children, helping her husband to achieve the goals of the family and being useful to the community in which she lives. But are all women satisfied with this role concept? Talented and ambitious women, in addition to these functions, want to develop their special aptitudes.
- Role as wife: As a wife, she is expected to be an affectionate companion, a good sex partner, confidant and social secretary of her husband. She has to take charge of the social life of the couple. She should develop interest in her husband's work. She should be able to understand his world of activities. Moreover she should be able to provide intellectual companionship to her husband.
- As a home maker: It is the duty of the woman to make her home a beautiful place to live in. She has to take care of the vital needs of the family such as nutrition, clothing, recreation etc.
- Role as a mother: She represents to her children the ideas and ideals of perfect womanhood working in harmony with her husband. The

child's first attachment is to his/her mother. She is his/her source of nourishment, warmth and comfort.

Changing Roles of Man and Woman Today

You may be wondering whether the couples are following the traditional roles of man and woman today. What are the changes that have occurred in their roles? Why? Let us look at the changed role concepts and the reasons for these changes.

Reasons for Role Changes

Today family roles are changing mainly because they have become less appropriate for the social and economic realities of the modern world. Mainly of the families are small in size and more women are employed. We can see changes in the styles of femininity as well as masculinity. The man is not maintaining the image of the brave, strong, tough, aggressive male of the past. The gentle, passive, submissive female is a character of yesterday. Opportunities are open for both sexes in education, work and family life. You may be interested to know what exactly the changes in role concepts are. Is there confusion in male's roles? We shall talk about these changes one through one.

- Sharing roles: Originally the husband was the head of the home and the chief power in the family. But now the domination of the man seems to be giving way to equalitarian roles flanked by husband and wife. More women have moved into employment outside the home. So the husband has to share the provider role with his wife. He also has to provide up his position as the custodian of family wealth as the wife has come to share spending. He has to accept her equality in community affairs also. They share household chores including the care of children.
- Economic equality: A working wife plays an significant role in decision making. When the women are employed, there is a tendency for the couple to share power equally. Wives are less dependent upon their husbands. The husbands in turn are freed from the burden of total support to their wives. Husbands give more emotional support to working wives.

Changes in the leadership role of man

The vocational demands may take the man absent from home for extensive periods of time. In his race for attainment and success, he may forget his major duty towards his family. Naturally we can find a decline in the respect given to husbands and fathers. The leadership role of man also tends to be reduced proportionately. At times, they stay absent from home to avoid responsibility in hard situations. Then, the handling of the situation falls on the shoulders of the mother. What happens when fathers are reluctant to take up their responsibilities? In such cases, children may develop sure behaviour troubles such as anxiety, aggressiveness and antagonism. You can observe the shrinking of father's role and an enlargement of the mother's role within the family. The masculine role has been greatly customized through the new role demands of the female.

Role expectations in dissimilar classes of society

Do you think that role concepts are the same in the middle of people of dissimilar social strata? There are differences. In the case of lower class males and females, the traditional definitions of masculinity and femininity persist. For them, roles are strictly segregated. But for the middle and upper classes, the spouses have sharing roles. They share responsibility in providing family finances, in bringing up children, and in giving them good education. Even in the middle and upper classes, is this kind of relationship accepted through everybody? The answer is no. In actual practice in India, the husbands have more say in the family.

Woman's expectations of man's role

Do you feel men are becoming more effeminate in today's culture? What do women expect from men? Men are taking more interest in domestic tasks, and in the care of children. But they are still less gentle, less family oriented and less domestically oriented than women.

There is no doubt that men are still trained to be more materialistic, and more success-minded than women. They have more efficiency, competence and material success. A women still expects several of these traits in a man. Several women still want the male to be strong and at times they want to depend upon males. They still admire and feel safe with a strong male. The man has to exhibit some courage and strength to meet the role expectations of his wife.

What do husbands expect from the wife?

The wife should become the companions of the husband. She is expected to provide love and affection to him. He expects an equal sharing of responsibilities. He expects cooperation, support and recognition for his efforts.

Factors Influencing Choice of Marital Roles

Marriage roles differ from several other roles in everyday life. There are no definite general role patterns. Each couple has to work out a role pattern for themselves. There are several factors influencing the choice of marital roles. Some of them are the following.

Family

Customary roles are learnt mainly from the family. It is through the attitudes, expectations and habits shaped in the family that a boy or girl gets vital training in role expectations for him or her as well as for the opposite sex. In some families the parents consciously instill what they feel is the appropriate sex role behaviour in their children. Children also learn sex roles and power patterns through imitating the behaviour of their parents.

The Peer Group

The peer group is oriented towards new attitudes and expectations for both sexes. They create new role expectations. These are usually in direct contrast to customary roles.

Culture

In our culture, men have more freedom of action. The women are expected to behave according to the norms dictated through society. Education has given her social and economic equality. But still she is restrained to traditional roles through our culture. This ties here down to home keeping and child care in addition to outside employment.

Employment of Women

This is a role assigned through culture, not through nature. Today, women are bearing a double load of work. Employment gives economic independence to the woman and she shares the financial burden of the family. She is fatigued after a day's work. There are more emotional and physical strains for the working woman. She may neglect her duties at home and may even refuse sexual rights to her husband. In some cases the husband may feel inadequate in his roles as provider. Children also suffer from lack of care and supervision. She may not be able to satisfy her role expectations. She may face the following troubles:

- Good health
- Willingness of the husband to share household duties
- Adequate pay
- A mother substitute if there are young children in the family and
- Co-operation from all family members A couple has to face several adjustmental troubles, if both of them are committed to their jobs.

ROLE CONFLICTS

You have already learnt that couples should have a clear concept of their roles when they marry. This concept shapes the foundation of marital adjustment. The goals in marriage can be achieved when these roles are played properly. Do you expect that the couple can adhere to role expectations always?

If not, why? When do role conflicts arise? Role conflicts occur when there are discrepancies in role expectations and the actual behaviour of the couple. It also occurs when partners relate to each other in some new or dissimilar ways because of emotional changes.

Reasons for Role Conflicts

No two persons are exactly alike. There are differences in attitudes, behaviour and beliefs. Husbands and wives gather dissimilar role concepts from their families. For success in marriage, each one has to produce in his/ her personality some resemblance of the partner's image of a man/woman. They have to change their values, attitudes and behaviour to fit in with the new role concepts. In the case of the woman living beside with her husband and other family members, the mother-in-law may criticize her. She may be forced to change her behaviour in order to fit in with her mother-in-law's concept of an ideal wife.

Role conflicts are brought in relation to by the strain put upon the wife in meeting the role expectations of her husband and his family and in adapting to the demands of his work. The husband also faces troubles when his wife is having an recognized role in her life. Money management is another area creating role conflicts. Money management is the handling and spending of the family income.

Usually conflicts arise in excess of the issues of who should handle money and how it should be spent. You may think it is a simple question touching only the surface. But the decision has far reaching effects on family life. It affects the attitudes and relationships of the couple. Troubles of domination, submission, insecurity, inferiority feelings etc. are created through the decision concerning the spending of family income. This in turn leads to conflicts flanked by husband and wife.

The other areas bringing in relation to the conflicts in marriage are unrealistic expectations from marriage, sexual incompatibilities, the discipline of children, thrash about for domination etc. Though, the mainly hard problem is hard in communication. The failure in communication occurs at a deeper stage of sharing feelings, expectations, intentions, and personal needs.

How can we resolve role conflicts?

In marriage, spouses are involved in the lives of each other. Hence disagreement in some areas is inevitable. How couple manage conflicts is more significant.

Disagreement Management

There are several ways in which conflicts can be supervised through the couple. In the first method, compatibility can be achieved when one partner is dominant and he/she attempts to get the other person comply with his/her desires and the partner agrees and accepts the complementary role. But when the disagreement is very strong and both partners are not willing to provide in, tension is intensified. They may emotionally withdraw from each other's relationships without finding out an actual solution for the problem. An uneasy peace is achieved. There may not be any quarrels, but both of them are not happy or satisfied. How can one manage conflicts without hurting the personalities of both the spouse? The best method to solve the problem in a reasonable way is through negotiating with each other. Changes are to be made in the roles of both partners. This helps them to learn and understand more in relation to themselves and in relation to the each other. This will deepen their relationship.

Personal Factors to be remembered

In quarrels, be very careful in the use of words. Pay attention to the good things your partner does. Do not pay too much attention to his/her faults. Don't fall asleep without making up a quarrel. Control you temper. Don't compare your spouse with your parents. Learn to compromise to forget and to forgive. The partner's should have flexible personalities. They should understand each other. Avoid scathing criticism and curb the impulse to make cutting retorts. Settle a problem through mutual discussion. If things cannot be resolved mutually, you will need to seek help from a family counselor or spiritual counselor. Although seeking help from a counselor is not common in India, we need to adapt ourselves to changing situations.

ROLE CHANGES IN DISSIMILAR STAGES OF MARRIAGE

As you have gone through the changes in marital roles you may have noticed that the relationship in marriage is not a static one. It is an ongoing procedure. In marriage you will find a variety of separate stages. Each stage requires new adjustments and re-valuation of the old ones. The couple has to re-organize their ideas, values, and goals for the smooth running of their married life.

Early years of marriage

Marriage takes the new couple to new and unfamiliar ways of life. Every day they have to take some decisions, solve some troubles and make plans for the future. The first year of marriage is one in which the husband and wife belong predominantly to each other, expenses are low, health is usually good and needs are simple.

Their unique plan of life is in the making. Each one is becoming recognized in the routine of his/her job. Throughout the first year or two of marriage the couple faces a multitude of adjustments. The following are some of the areas of adjustments in marriage which will be affecting their behaviour or role in early marital life.

Relationship flanked by husband and wife

When you get married you say "I take you". It is not presently the company, thoughts or the body of the partner that you take. You take the whole of the other person.

Accepting each other

The husband and the wife have to accept each other with all their shortcomings. At the same time, they should uphold their individually, personal worth and self-respect.

Communication

The spouses should show the ability and willingness to communicate with each other. Communication is the breath of married life. It is not enough to love, to care or to respect. The lover necessity expresses affection.

Provide and take

The couple should possess the ability to provide and take. To provide oneself to the other means to share one's mainly intimate feelings. They should be co-operative and should be capable of emotional interdependence.

Personality factors in relation to role expectation

You know man and woman differ in their physical structure. Do they show differences in their emotional and psychological structure too? As every cell of the body is dissimilar, so also every feeling, every reaction, and every attitude of man and woman have scrupulous features according to one's sex.

The respective nature of man and woman determines their roles in marriage. Success in marriage can be attained when the couple respects the feature qualities which nature has given to each of them and when they adhere to their respective roles. Do not expect one's own features in the other partner and do not try to change him/her.

Personal habits

Mainly of the troubles faced through the couples arise from their vital personality patterns. Each person learns scrupulous habits and attitudes from early childhood.

So regardless of their similarities, each partner, always brings to marriage enough differences in their attitudes, punctuality, formality and informality in housekeeping and management of home, observing social conventions etc.

may bring in relation to the role conflicts in marriage. Only if both partners are willing to compromise from the beginning of marriage, can they get beside smoothly.

Sexual relations

Why do you consider sex as an significant factor for the success of marriage? It is because; sexual relationships directly and indirectly power the adjustment in other areas of life. It is the physical foundation of marital relationships. If there is any frustration in sexual relationship it will affect the emotional, mental and social adjustment of the concerned person.

Sexual adjustment in marriage

You know that biological factors are significant in determining the sexual satisfaction of the individual. Cultural factors are also equally significant. What are the other factors?

From where does an individual gain his sexual attitudes? The sexual attitudes and expectations are learnt through an individual from his/her early childhood from his/her family and his peer group.

The cultural norms of the society in which he/she grows up and matures and his/her biological structure vary the acquired attitudes of the person. Knowledge of sexual anatomy and appropriate sexual activities are also necessary for good adjustment.

The other factors conducive for good sexual relationship are complete emotional acceptance of each other and an understanding of each other's behaviour.

Sexual maladjustments in marriage

Unfavorable attitudes towards sex, ignorance in relation to the sex and sexuality lack of proper sex education, unfavorable sexual experiences of early years of life, unrealistic expectations in relation to the sex in marriage and the unwillingness to be open with one's partner etc. are some of the reasons for sexual maladjustment. Psycho sexual development powers an individual's sex life. Fixation or regression at any stage of development will adversely affect the sexual adjustment.

According to Freud at one stage of development, the male and female children are attracted to the parent of the opposite sex – the boy to the image of his mother, the girls to the image of her father. For the boy this phenomenon is recognized as Oedipus intricate, for the girl it is Electra intricate.

Normally the boy and girl will pass through this stage and begin to identify with the parent of one's own sex. If they are fixated to this stage, in their later marital life it will cause troubles in sexual relationship with the partner.

Confusion in relation to the sex and false attitudes towards sex

In Indian society discussion on sex continue to be a taboo since extensive. This conservative attitude of elders prevents the imparting of healthy facts in relation to the sex to the teenagers. So they may turn to some other source to get information. They may be knocking at the wrong doors and getting wrong and dangerous information. Inaccurate and distorted information creates unhealthy attitudes towards sex. Later on in married life it will create a number of troubles. The role of man and woman in sexual adjustment depends upon the attitudes and expectations each person brings into marriage. Both the spouses should have balanced and positive attitudes towards sex.

Relations with in-laws

Marriage joins not only the man and woman together, but their families also. Hence, in-laws are the new relatives acquired through marriage. You may wonder whether in-laws are bringing troubles to every marriage. The relationship with in-laws may bring in relation to the problems in some marriages.

The circumstance in which each person lives is dissimilar. If the son-in-law or daughter in- law fits in with the expectations of the respective family he/she will be accepted easily. Otherwise troubles may arise.

Interference of mothers

Usually it is the mother-in-law who is more involved in in-law troubles. Mother's lives are tied up with the lives of their children. They are very close to children. Some mothers are reluctant to provide up their roles and try to uphold their responsibilities and expect privileges from their married children. The boy or the husband should understand this phenomenon and manage the situation accordingly.

Husband's mother interferes more often

The mother is the first woman in the life of her son. After the son's marriage some mothers cannot accept another woman loving and caring him, even though it is his wife. The wife feels the mother-in-law to be a threat to her marital happiness. Only an intelligent and understanding husband can solve this problem. He should provide due importance to both of them. Don't make them rivals. The wife should not be in competition with the mother. Provide mothers time to find out new interests.

Other in-laws

Sister-in-law also creates troubles in the life of married couple. Brothers-in-law and fathers-in-law also share a small part in the in-law troubles. Care of elderly relatives also creates troubles.

Why do the parents interfere?

Parents are concerned with the growth and achievements of their children from the time of their birth. Hence, when children get married it is hard for the parents to ignore or to be indifferent in relation to the lives. At the same time the young couple is eager to be self-governing. They do not like the interference of parents. In several cases, the parents are not economically self-governing. They are forced to live with their married children. This also creates troubles.

Positive contribution of in-laws

The in-law relationships are not always a problem. If the young wife appreciates her husband's mother and loves her as a second mother, she will find her very helpful.

Cultural factors affecting role changes

The role expectations vary with dissimilar cultures. Culture circumstances the personality of an individual. Marital roles are determined through the culture in which one grows up. The couple coming from two dissimilar cultures will behave differently in a given situation.

They will be having dissimilar expectations and their goals in marriages will be dissimilar. There may be few common interests. Each brings from his/her culture quite often dissimilar values, attitudes, customs and styles of living. Cultural differences make mutual understanding and communication hard. What should the couple do to adjust to each other?

The couple requires more adjustment than others in a homogenous marriage. Before marriage, they should think cautiously whether they are capable of making all the adjustments in such a marriage. To achieve marital satisfaction they should work with more maturity, understanding and determination than a couple marrying from similar backgrounds. If the couple can accept one another without trying to recast the mate into one's own scrupulous mould, they may get beside well.

Religion

Society may accept inter-racial marriages. But it seldom approves inter-religious marriages. The chances for success of an interfaith marriage depend upon the ability of the couple to face the situation maturely and realistically. Before marriage, they should have a clear evaluation of the difficulties they have to encounter and overcome.

Areas of disagreement

One's desire to make the other person adopt his/her religion causes major marital troubles. Whether the parents and friends will accept the match is another difficulty. This affects the relationship of the parents. The religion to which children shall belong their education, choice of child training methods, the use of birth control events etc. are some of the major areas of disagreement.

Adjustments

If they are of an understanding nature and tolerant of each other's religion and have no desire to impose their own religion on the mate, they will get beside well. If they have a clear thought of the several obstacles they have to encounter and if they are flexible enough to make their adjustments, their marriage may became successful.

Importance of early years

Throughout the early years the couple has to make all the above adjustments for the success of their marriage. Each partner necessity learns how far he/she can take the mate in dissimilar areas and which the things he/she necessity avoid are. A couple may be able to make adjustments successfully in the early years. But marital life is a continuous procedure. Aging and maturity bring in relation to the changes in the lives of the couple. The new and changing situations in life makes further adjustments in the roles of husbands and wives necessary and inevitable.

Role changes with the arrival of children

It signals the beginning of a new stage in the relationship of husband and wife.

Life beside with offspring

There are radical role changes to be made with the birth of children. Parenthood brings fulfillment and completion to marriage. It fundamentally changes the whole character of marital relationships.

Troubles related to transition to parenthood

The coming of the child is early anticipated, loved and cherished through the parents. But transition to parenthood brings in relation to the number of readjustments in the lives of parents. The arrival of the child upsets the balance maintained in the family.

Parental role

Parental role brings personal, social and economic privations.

Economic troubles

The father is worried in relation to his responsibility and additional expenses. The wife may have to provide up her career and take up the role of a full time housewife. This affects the income and expenditure of the family.

Personal difficulties

The wife may be embarrassed in relation to the circumstances of pregnancy and its effects on her social life and marital relations. Men also have to make role changes. When the wife is overburdened with child care and other household chores, he may have to share her duties.

Social participation

The social life and recreation of the parents are severely affected. The young couple is tied down with the care of children. When they can entrust some one else for child care, they can have joint participation in social activities. Otherwise they attend social functions separately.

Though, joint participation gives more satisfaction to the couple. You have already learnt that, in the early stage of marriage, the couple had a child-centered relationship. They had very few outside interests. Their lives were built approximately their children.

Feelings of uselessness

Some women feel that their lives are empty and futile. When parental responsibilities diminish, they may be having a feeling of uselessness. For such women, ending of parental role is a traumatic experience. It may cause neurotic difficulties. They can try to learn new skills and become occupied in some activities. Husbands are occupied with their work. In this way, they can adjust to the changes in the family.

Adjustment with in-laws

Children may get married and the parents have to adjust to the children's spouses. Another adjustment is the care of aging parents. Caring for the elderly may help to fill the gap created when children leave home. But in several cases the care of aging parents deprives the couple of the opportunities for social participation and development of new interests.

Role as grandparents

You may be well aware that several men and women become grandparent before grand parent middle age ends. This is a new role in the life of the couple.

Formal roles

Some grandparents play a formal role following a hands off policy as far as care and discipline of grandchildren are concerned.

Surrogate parent role

In some other cases they assume the responsibility for the care of children. Grandmothers are more active in this role.

Fun seeker role

Another role is that they impart special knowledge to their grand children and teach them sure skills. The role of grand parents gives couple more satisfaction than their earlier role of parents.

Role changes in old age

You necessity have observed the role changes in early adulthood and middle ages of the couple. What are the changes throughout old age? How does the couple adjust to these changes? You also know that the couple had a close, intimate relationship in the early stage of their marriage. Later on, it changed into a child-centered relationship. In old age, the relationship flanked by a husband and wife again takes the form of a pair centered relationship.

Effects of retirement

The pattern of family life undergoes further changes with the retirement of the husband. Perhaps the greatest adjustment of all occurs throughout this period. The husband retires and suddenly faces the insults of old age. He feels unwanted and has no motivation to compete with other men. He feels lost and does not know what to do with his free time. He tends to be depressed and unhappy. He is irritable in his treatment of his wife. He always criticizes and finds fault with the spouse. He does not want to help the wife in her household duties thinking it is woman's work.

Common interests

Retirement forces the couple to be together mainly of the time. They can develop a pattern of togetherness in recreational activities. The same pattern can be applied to other areas of life. With the change of time mutual interests can be urbanized. When children leave home the parents are drawn towards each other more closely.

5

Policies and Programmes

INDIAN FAMILY IN TRANSITION

FAMILY STRUCTURE, FUNCTIONS, AND RELATIONSHIPS

Family may be defined as a group of persons united through the ties of marriage, blood or adoption or consensual unions, usually constituting a single household, interacting and communicating with each other and creating and maintaining a common culture.

Family Functions

There are three major areas of functions which have been particularly accentuated. Thus a functional analysis of the family emphasizes the relationship flanked by the family and the superior society, the internal relationship flanked by the sub-systems of the family, and the relationship flanked by family and the personality of individual members. The former may be termed macro functionalism and the latter two micro-functionalisms.

Family Patterns

Family patterns are conceptualized in terms of family composition. A household is one of the dimensions of the family pattern. It is a residential and domestic unit composed of one or more persons living under the same roof and eating food cooked in a single kitchen.

The normative family patterns in India are extended or joint family and elementary or nuclear family. The term joint family is used more commonly in India than extended family.

The joint family comprises movable or immovable property, and all the members of the family may or may not be staying together.

The elementary or nuclear family comprises couples and their unmarried children and is usually financially self-governing of other families.

Family Structure

Family structure is conceptualized as the configuration of role, power and status and relationships in the family. It depends upon the family's socio-economic background, family pattern, and extent of urbanization. Family structure has implications for family unity and stability and development of individuals.

The functions that the family as an institution performs are divided in the middle of family members in the form of roles. Roles are culturally defined and are passed on to succeeding generations as correct behaviour. Role expectations are thus learnt from family orientation. Norms of family relationships may comprise norms and dyadic relationships, families and kinship orientation. The dyadic relationships cover the following dyads in the family:

- Filial Relation: Parent-child relation
- Fraternal Relation: Relation in the middle of siblings
- Conjugal Relations: Relation flanked by husband and wife
- In-law Relation: Relation flanked by family members related through marriage and not through blood.

Family Practices

Family practices depend upon the family's ethnic background, extent of urbanization, family structure, and family laws. Marriage practices cover marriage patterns, selection of marriage partner, and age at marriage, age at consummation of marriage, marriage rituals, financial swap, and divorce. Other family practices relate to lineage, residence, child bearing, child birth, adoption, guardianship, and custody of children, maintenance, death and inheritance and succession.

Families are classified into patrilineal and matrilineal families according to the lineage or descent through the father or through the mother. Patrilineal families are usually patrilocal and matrilineal families are usually matrilocal through residence. Neolocal families establish a new residence after marriage.

Role of the Family in the Socialization of the Child

Socialization is the procedure through which persons learn the ways of a given society. Children are taught the ways and values of their society through get in touch with already socialized individuals.

The family is significant because it maintains get in touch with children in excess of a longer period of time than any other group, and through close emotional association is able to exert maximum power in their lives. Both adult and other sibling members (brother and sisters) act as role models in providing examples for children in the development of their personality, attitudes, and behaviour.

Family Dynamics

Family dynamics broadly comprise family interactions and family development as influenced through socialization of family norms of individual members.

Family Interactions

Family interactions refer to the distinctive character of interpersonal relationships which occur flanked by members of the family. An interpersonal relationship is a relation based on personal interaction rather than on any legal or structural foundation. The main features of family interactions may comprise cohesion, communication, role performance, decision-making, and adaptability. Each subsystem of the family has dissimilar interactional patterns.

Family Cohesion

Family cohesion is defined as the emotional bonding that family members have towards one another. There are four stages of cohesion, ranging from disengaged (very low) to separated (low to moderate), to connect (moderate to high), to enmeshed (very high). It is hypothesized that the central stages of cohesion (separated and linked) make for optimal family functioning. The extreme stages (disengaged and enmeshed) are usually seen as problematic.

Family Communication

Family communication is defined as all the verbal and non-verbal behaviour that occurs within the family, and flanked by the family and its social environment.

Role Performance

The following are the dissimilar characteristics of role performance:

- Role enactment, role performance and role behaviour
- The behaviour dimension of roles either validates the cultural expectations or emerges to create new roles.

Role Commitment

Role competence is the evaluation of one's own and another's performance.

Role Disagreement

Inter-role disagreement takes place when the norms or behaviour patterns of one role are inconsistent with those of another role of the same individual. Intra-role disagreement takes place when two or more categories of people hold conflicting expectations concerning the behaviour appropriate to a single role.

Decision-making

Decision making involves recognition of need for decision, identification, and weighing of acceptable alternatives, selection of an alternative and facilitating its action.

Family Adaptability

Family adaptability is defined as the ability of a family system to change its power structure, role relationships, and relationship rules in response to situational and developmental stress. It is hypothesized that the central stages of adaptability (structured and flexible) are more conducive to marital and family functioning with the extremes (rigid and chaotic) being the mainly problematic.

Family Development

The concept of family development gives a longitudinal view to compare family interactions in dissimilar stages of family life span. These stages may be determined through the age and developmental needs of the adults and the children in the family. The Family Life Cycle is a way of taking look at the family life. It is based upon the recognition of the successive patterns within the stability of family living in excess of the years. Evelyn Duvall depicts the family life cycle as consisting of 8 stages.

- Stage I Beginning families (married couple without children)
- Stage II Childbearing Families (oldest child, birth to 30 months)
- Stage III Families with Preschool Children (oldest child 2½ to 6 years
- Stage IV Families with School Children (oldest child 6 to 13 years)
- Stage V Families with teenagers (oldest child 13 to 20 years)
- Stage VI Families as launching centers (1st child gone to last child's leaving home)
- Stage VII Families in the middle years (empty nest to retirement)
- Stage VIII Aging Families (retirement to death of one or both spouses)

Development Tasks

A family's development task is a growth responsibility that arises at a sure stage in the life of a family, successful attainment of which leads to satisfaction and success with later tasks, while failure leads to unhappiness in the family, disapproval through society, and difficulty with later developmental tasks. Family development tasks are vital family tasks specified for a given stage to development in the family life cycle. Family life development programmes aim at development of knowledge, attitudes, and skills towards democratic family functioning and strong family ecology. These goals may be achieved through the following tasks:

- Attitude development in favor of family rights and responsibilities.
- Social training for enrichment of family dynamics and development

at each stage of family life span for strengthening family's interactions with its social ecology.
- Information dissemination in relation to the family possessions such as laws, policies and implementation systems and services.

SOCIAL CHANGES AFFECTING INDIAN FAMILY

The family has been and continues to be one of the mainly significant elements in the fabric of Indian society. The bond that ties the individual to his family, the range of the power and power that the family exercises make the family in India not merely an institutional structure of our society, but accord provide it a deep value. The family has indeed contributed to the stability to Indian society and culture.

Today, the Indian family is subjected to the effects of changes that have been taking place in the economic, political, social, and cultural spheres of the society. In the economic sphere, the patterns of production, sharing, and consumption have changed greatly. The procedure of industrialization and the consequent urbanization and commercialization have had drastic impacts on the family. Migration to urban areas, growth of slums, change from caste oriented and hereditary occupations to new patterns of employment offered through a technological revolution, the cut-throat competition for economic survival and several other economic changes have left their impact on the family.

Briefly speaking, these changes in the socio-economic political- cultural milieu of our society have led to changes in the structures, functions, roles, relationships and values of the family. In the context of the changes in the economic system, more and more members of the family are moving absent from the superior family circle and living as individuals or members of a nuclear unit in urban areas. The patterns or loyalties, obligations and expectations have changed. The case of the child and the aged in scrupulous have become a problem for several due to structural changes in the family.

Change in Traditional Functions

Several functions performed through the traditional family are being taken in excess of through other agencies such as schools, day care centers, commercial and entertainment centers, etc. For instance, a significant function of cultural transmission performed through the family has been affected because the nuclear families are scattered cultures. The function of setting moral standards for the rising children and adolescents has been taken in excess of to a large extent through the peer group culture, mass media or through commercial entertainment. A major area of the family that has been affected through the social and economic changes in the society is that of the role performed or expected of the dissimilar members of the family. The traditional role allocation based on sex, age, or kinship has changed. With more and more women taking up jobs outside the home, the traditional role of the wife has changed. Likewise the role of the father, the mother, the husband, the child, and the elders in the family has all undergone several changes. The degree of role-overlap depends on the version of the scrupulous family to the changed situations.

The changes in roles have inevitable affected the relationships, in the middle of the members of the family. The concept of freedom, individuality, and rights of the individual has had their impact on the relationships too. The attitudes of implicit obedience to elders, concern for others, self-denial for the sake of others in the family, acceptances of the power of parents and superior status of the male are being displaced through attitudes of self-centeredness, assertion of individual rights, clamor for equality and right for self-determination etc. In the realm of values, today's family is moving towards materialism, individualism, and liberalism. The cherished values such as respect for age, concern for the weak, devotion to one's duty, co-operation are being replaced through competition and "getting ahead". It is not surprising that the family in India is also succumbing to the pressures of the time.

The consequences of these changes are several. Troubles such as child neglect, behaviour troubles in children, indiscipline in the middle of the youth, alcoholism, drug addiction, neglect of the elderly, material disharmony etc. are on the augment today and are indications that the family is not able to handle the change in a desirable manner. There is need so, to help the families to cope with the pressures and challenges of their life situation which are affected through the interacting forces to change in the economic, political, and cultural spheres.

Alternative Family Patterns

One of the mainly striking characteristics of contemporary societies is the attendance of a range of family variations, from the mainly traditional, extended families with strict, gender based sex roles to the modern dual career families based on liberal, equal sex roles and to adults cohabiting without marriage. The term "alternative family patterns" suggests family patterns that result from personal circumstances outside one's control (death of a partner, infertility) or from socio-economic circumstances (male migration, work participation of women). In the Indian context, mainly family variations are a result of personal or socio-economic circumstances. Experimental or chosen lifestyles like living without marrying, and being childless voluntarily are restricted to a very small group of people. The following are the mainly commonly observed family variations in India.

- Single parent families
- Female headed households
- Dual earner/career families

- Childless families
- Adoptive families

TROUBLES ENCOUNTERED IN THE FAMILY AND INTERVENTION PROGRAMMES

Programmes

The following is a list of major family problem situations which may be entry points for family intervention. The causal factors of the situations may be multiple, situated in the family ecology, socialization of authoritarian family norms, dysfunctional family interaction patterns, or troubles with individual member's developmental tasks. These areas are, so, not mutually exclusive. Some of these problem situations are:

- Families with individuals having troubles:
- Families of the disabled
- Families of the chronically/terminally ill
- Families of substance abuse addicts

Troubles with child bearing:

- Infertility
- Unwed motherhood

Marital troubles:

- Marital disharmony
- Marital breakdown

Abuse and violence in families:

- Child abuse in family
- Family violence against women
- Elderly abuse in family
- Family abuse of the disabled

Families in disagreement with other systems:

- Families with unemployment/indebtedness
- Families with inadequate or no land/housing

Families affected through dis-equilibrium in other systems:

- Families facing political violence
- Families facing environmental disasters
- Uprooted/refuge/migrant families

Family deprivation:

- Destitute children
- Destitute adults
- Destitute aged

These family situations, in turn, may affect the family functions, interactions, and individual members. The family may become incapable of carrying out some individual members. The family interaction patterns may change as a coping mechanism. Harm may be caused to physical and mental development and health of individual members, particularly children, women, and the aged. In the extreme situation family may disintegrate and individual members may become destitute.

Family Intervention

A family practitioner's role comprises planning and implementing a combination of interventions ranging from development to remedial, by individual, group, as well as community methods to strengthen these families, coping mechanisms and then rehabilitate them. Family counseling, marital counseling, family, and marital therapy crisis intervention, encouraging self help groups and legal aid are specific methods that may be used. The groups needing family intervention may be children, adolescent, youth, women, couples, or the aged; but the family as a whole may be measured the unit for intervention. Besides, planning and implementing the services, the practitioner needs to monitor and evaluate them and raise public awareness in relation to these services.

GOVERNMENT POLICIES AND PROGRAMMES FOR FAMILIES

Family policy means everything that the government does for the family explicitly or implicitly. The Constitution of India does not make an explicit reference to the family. Though, it lays down the fundamental rights of individual citizens in terms of equality, nondiscrimination, and protection. Its Directive Principles of State Policy state that, "The state shall strive to promote the welfare of the people through securing and protecting as effectively as it may a social order in which justice, social, economic, and political shall inform all the institutions of the national life". Article 41 specifically lays down that, "the state shall, within the limits of its economic capability and development, make effective provision for securing the right to work to education and to public assistance in cases of unemployment, old age, sickness, and disablement".

Social Policies and the Family

Reviewing the social policies of India that affect the family, one can conclude that India has explicit family policy with explicit goal for the family only in the form of the Family Welfare Programme of the Ministry of Health and Family Welfare. This programme pursues family planning as an integral part of a comprehensive policy, covering the total health care delivery system. Though, the policy has the limited goal of promoting Planned Parenthood with a two children/one child norm, through the self-governing choice of family welfare methods. We have a National Policy for Children that aim at full physical, mental, and social development of children. As far as the family is concerned, it states that, "In organizing services for children efforts would be directed to strengthen family ties so that full potentialities of growth of children are realized within the normal family, neighborhood, and community environment". In 1992, India adopted a "National Plan of Action" for children, based on the recommendations of the World Summit of Children contained in the plan of Action for Survival, Protection, and Development of Children. The extensive-term goals of the National Housing Policy are eradication of home—— lessens improvement of the housing circumstances of the inadequately housed, and provision of minimum stage of vital services and amenities to all.

The National Policy for Child Labour, the National Youth Policy, the National Policy Education, and the National Health Policy aim at specific services for select groups of individuals and implications for the family are implicit. India has a forest policy which aims at protection, conservation, and development of forests. Thus, we have several policies for the family and its members. A draft of the national policy for the aging is being discussed. A National Policy for Women is yet at the stage of recommendations of the National Perspective Plan for Women. Our social security legislation is applicable only to work force in the organized sector. A large number of families at risk remain uncovered through any social security against poverty, old age, disability, illness, and environmental disasters.

Family Laws in India

India has dissimilar personal laws for families belonging to dissimilar religions, and so, do not have any agreed upon goals in relation to the significant aspect of civic life. Hindus, Muslims, Christians, Jews and Parsis, have their own personal laws, which cover matters of personal relations and family practices such as marriage, divorce, adoption, maintenance, guardianship and custody of children, and inheritance and succession.

As these laws draw from the respective religious norms, they often perpetuate traditional patriarchal norms and slow down the procedure of reforms. Although Article 44 of the Constitution maintains that, "The State shall endeavour to secure for the citizens a uniform civil code throughout the territory of India", very few attempts have succeeded in enacting secular family laws that apply to family practices of all Indians.

These secular family laws are, the Child Marriage Restraint Act, 1929. The Medical Termination of Pregnancy Act, 1972, Special Marriage Act, 1974, Dowry Prohibition Act, 1961 and provisions made in the Criminal Procedure Code and Indian Penal Code.

Government Schemes and the Family

In the absence of an overall family policy, the government schemes that affect/benefit families and their members are divided in the middle of dissimilar ministries.

Provisions under Ministry of Welfare

The Ministry of Welfare has five bureaus for the following purposes. The Bureau of Social Defense controls the troubles of family and social disorganization which are manifest in the shapes of delinquency, juvenile vagrancy, drug addiction, alcoholism, and other such troubles of personal and social deviance, within the framework of specific laws and allied events. This Bureau implements the following schemes.

- For the welfare of children in need of care and protection;
- For the prevention and control of juvenile special maladjustment;
- For spreading awareness in relation to the ill effects of drug abuse and providing counseling, de-addiction, after-care and rehabilitation services;
- For refugees from other South Asian countries (Relief and Rehabilitation schemes);
- For grant-in-aid to welfare programmes of the aged; and
- A scheme for the development of street children.

Provisions under Ministry of Human Possessions Development

The Ministry of Human Resource Development runs the Department of Women and Child Development which has two bureaus: Bureau of Nutrition and Child Development and Bureau of Women's Welfare and Development. The Department runs the following schemes for children:

- Integrated Child Development Services for Children;
- The Adolescent Girls Scheme;
- A Scheme of Crèches/Day Care Centers for Children of Poor, working and ailing women and;
- Early Childhood Education Programme.

The Department runs the following schemes for women:

- Hostels for Working Women Belonging to Low Income Group;
- The Scheme for Employment cum Income Generating Production Units for Women;
- The Scheme for Training Centers for Rehabilitation of Women in Distress;
- The Scheme for Short Stay Homes for Women and Girls;
- Public education for the Prevention of Atrocities against Women; and
- The Support to Training cum Employment Programme (STEP)

The Central Social Welfare Board, which is an autonomous organization working under the Department of Women and Child Development, offers the following schemes:

- The scheme of Opportunities for Education For Adult Women;
- The Socio-Economic Programme provides opportunities for work and wages to needy women that contain the economically backward, the

destitute, the widowed, the deserted, the physically handicapped, and so on;
- The Awareness Generation Project for Rural and Poor Women Programme;
- Mahila Mandals;
- Family Counseling Centers;
- Voluntary Action Bureaus;
- Welfare Extension Project; and
- Scheme of Balika Mandals for Adolescent Girls.

Provisions under the Ministry of Urban Development

The following Social Housing Schemes are implemented through the Ministry of Urban Development:
- Housing Scheme for economically weaker sections;
- Housing Scheme for low income groups;
- Housing Scheme for middle income groups; and
- Rural housing sites cum construction assistance scheme for landless workers;

The Ministry is also implementing the following programmes related to urban poverty alleviation:
- The Nehru Rozgar Yojana;
- The urban vital services for the poor; and
- Environmental improvement of urban slums aims at ameliorating the living condition of urban slum dwellers.

Provisions under the Ministry of Rural Development

The Ministry of Rural Development implements several schemes. Though, with the formation of every new Ministry and Government at the centre, each of the programmes gets a new name or is put into dissimilar categories. Some of the well recognized programmes contain:
- Integrated Rural Development Programme (IRDP)
- National Scheme for Training of Rural Youth for Self-Employment (TRYSEM); and
- Development of Women and Children in Rural Areas (DWCRA).

HUMAN RIGHTS IN THE FAMILY

The goal of the International Year for the Family (IYF), 1994 is "Building the smallest democracy at the heart of the society". The plans for the IYF seek to promote the vital human rights and fundamental freedoms accorded to all individuals through the set of internationally agreed instruments formulated under the aegis of the United Nations, whatever the status of each individual within the family, and whatever the form and condition of the family.

According to the United Nations, "Human rights could be usually defined as those rights which are inherent to our nature and without which we cannot live as human beings. Human rights and fundamental freedoms allow us to fully develop and use our human qualities, our intelligence, our talents and our spiritual and other needs", their being met is, so, not a matter of choice, but an imperative of vital justice. The human rights need to be applied to the family for enriching family life. Denial of them creates circumstances of use, deprivation, and destitution of families and their individual members. Family responsibilities are as significant as family rights to ensure family happiness. The following human rights instruments are applicable to the family:

- The U.N. Declaration of Human Rights;
- The U.N. Convention on the Elimination of All Shapes of Discrimination against Women; and
- The U.N. Convention of the Rights of the Child.

Desai has applied these human rights instruments to family and added family responsibilities at three stages:

- The individual's right to have a family;
- The individual's rights and responsibilities within the family, and
- The family's rights and responsibilities with reference to its environment. These family rights and responsibilities can form the goals for family well-being that cut crossways the diverse shapes of families.

The Individual's Right to Have a Family

- Every child has the right to be reached through his or her natural family. Parents have the primary responsibility to rear their children.
- Every adult has the right to marry and form a family
- Every elderly person has the right to be cared through his or her children.

The Individual's Rights and Responsibilities within the Family

There are natural humane and caring qualities in family relationships. These may be used and strengthened through promoting and protecting every individual's rights for status, worth and dignity; equality and nondiscrimination; freedom and choices in family life; social security from family members; and protection from family abuse and violence. It is the responsibility of every individual, family, community and the state to promote and protect these rights. Every individual has responsibility to enrich family interactions. These rights and responsibilities are further elaborated below.

- Every family member is equal in dignity and worth, irrespective of age and gender. Every family member, so, has the right to equal allocation of family possessions, and equal responsibility towards household work.

- Both the married partners have the right to gender equality in marriage with respect to monogamy, role, power and status; parenting, guardianship and custody of children; title to matrimonial home and property; and dissolution of marriage and division of matrimonial property.
- Every family member has the right to freedom and choice in family life.
- Every family member has the right to care and support from other family members in crisis events such as disability, illness and in old age.
- Every family member has the right to life and security of persons.
- Individuals need legal protection of the state for their rights within the family.
- It is the responsibility of every family to register birth, marriage, and death of every family member.
- It is the responsibility of every family member to promote sensitivity and responsiveness, positive communication patterns, democratic decision making and peaceful and nonviolent approaches for resolving conflicts in their family interactions.

PLANNING FOR FAMILY AND FOR RESPONSIBLE PARENTHOOD

A marriage undergoes a transformation when husband and wife suddenly become parents. The word "suddenly" is appropriate because the transition to the status of parent is quit unlike the transitions to other major statuses in life, such as marriage or an occupation. People have relatively greater preparation and experience for entering marriage or an occupation, and they also have a grace period throughout which they slowly assume the responsibilities of the new status. The transition to parenthood is quite dissimilar and often dramatic. It may constitute a crisis in the life of married couple, as it forces them to take on a significant and demanding responsibility. The task of parenting allows the wife and husband much less time to devote to each other. And as the child becomes an interacting individual in the family, intricate relationship pattern emerge.

Once the motivation for the good life is created, couples will be able to come to mutual decision concerning each pregnancy so that every child in the family will be a 'wanted child' and not a "product of chance". Every pregnancy will place husband and wife in a new human setting and far from experiencing pregnancy as a pathological condition. It will place them in a revived experience at the very centre of human destiny. The decision to marry and to bring a child into the world is the couple's decision. They necessity responsibly prepare for parenthood so as to give for themselves and their children in an environment in which they can grow personally and as a family. Marriage and conjugal love are through their nature ordained towards the begetting and educating of children. Children are really the supreme gift of marriage and contribute very substantially to the welfare of their parents. Hence conjugal love requires in husband and wife an awareness of their mission of "responsible parenthood".

Characteristics of Responsible Parenthood

- Responsible parenthood means the knowledge and respect of their functions.
- The deliberate and generous decision to raise a family, or through the decision, made for some motives and with due respect for the moral law, to avoid for the time being or even for an indeterminate period, a new birth.
- The responsible exercise of parenthood implies that husband and wife recognize fully their own duties towards God, towards themselves, towards the family and towards the society, in a correct hierarchy of values.

Future Trends

The future of parenthood will contain a greater awareness of what the role involves, increased sharing through both spouses of the birth of their child or children, new techniques of delivery, and more emphasis on fatherhood. There is need for preparation for parenthood classes accessible to potential parents as Lamaze classes (the Lamaze method emphasizes active involvement of the father in the delivery procedure).

FAMILY PLANNING POLICIES

OBJECTIVES AND SCOPE OF FAMILY PLANNING

An Expert Committee of the WHO defined Family Planning as: a way of thinking and living that is adopted voluntarily, upon the foundation of knowledge, attitudes, and responsible decision through individuals and couples, in order to promote the health and welfare of the family group and thus contribute effectively to the social development of a country".

Objectives of Family Planning

Family planning refers to practices that help individuals or couples to attain sure objectives:

- To avoid unwanted births
- To bring in relation to the wanted births
- To regulate the intervals flanked by pregnancies
- To control the time at which births occur in relation to the ages of the parent and

- To determine the number of children in the family

Now you have learnt the definition and objectives of family planning. Let us see the scope of family planning services.

Scope of Family Planning Services

It is not synonymous with birth control, but is in fact more than mere birth control. A WHO Expert Committee has stated that, family planning comprises in its purview.

- The proper spacing and limitation of births
- Advice on sterility,
- Education for parenthood
- Sex education,
- Screening for pathological circumstances related to reproductive system,
- Genetic counseling
- Premarital consultation and examination
- Marriage counseling,
- Carrying out pregnancy tests,
- Preparation of couples for the arrival of their first child,
- Providing services for unmarried mothers.
- Teaching home economics and nutrition and,
- Providing adoption services.

These activities vary from country to country according to national objectives and policies with family planning. This is the modern concept of family planning. Rapid population growth in less urbanized countries is a key factor in limiting the ability of these countries to raise standards of living.

Significant obstacles to their socio-economic development contain limited possessions, food sharing troubles high rate of diseases and infant mortality, lack of proper sanitation, scarcity of funds and shortage of educational facilities and work opportunities. In this context, the Planning Commission clearly, recognized the need for population control right at the beginning of the planning exercise.

To quote the First Five Year Plan:

- "The recent augment in the population of India and the pressure exercised on the limited possessions of the country have brought to the forefront the urgency of the problem of family planning and population control. It is, so, apparent that, population control can be achieved only through the reduction of the birth-rate to the extent necessary to 'stabilize the population' at a stage constant with the requirements of national economy. This can be secured only through the realization of the need for family limitation on wider scale through the people".

NATIONAL POPULATION POLICY 2000

A social policy signifies consensual social purpose, and aims at progressive and structural changes. It pays appropriate attention to economic-cultural, political, and social factors as also to short-term and extensive term perspectives. A population policy can be nothing less than a social policy. Population programme necessity works itself in the whole fabric of social environment and necessity power and be influenced through all other events of social changes. When a policy is translated into programmes and activities, it causes social development, with due involvement of integrated diverse range of sectoral programmes and activities.

National Population Policy 2000

- Address the unmet needs for vital reproductive and child health services, supplies and infrastructure.
- Make school education up to age 14 free and compulsory, and reduce drop outs at primary and secondary school stages to below 20 percent for both boys and girls.
- Reduce infant mortality rate to below 30 per 1000 live births.
- Reduce maternal mortality ratio to below 100 per 100,000 live births.
- Achieve universal immunization of children against all vaccine preventable diseases.
- Promote delayed marriage for girls, not earlier than age 18 and preferably after 20 years of age.
- Achieve 80 percent institutional deliveries and 100 percent deliveries through trained persons.
- Achieve universal access to information/ counseling, and services for fertility regulation and contraception with a wide basket of choices.
- Achieve 100 per cent registration of births, deaths, marriage, and pregnancy.
- Contain the spread of Acquired Immunodeficiency Syndrome (AillS), and promote greater integration flanked by the management of reproductive tract infections (RTI) and sexually transmitted infections (STI) and the National AIDS Control Organisation.
- Prevent and control communicable diseases.
- Integrate Indian Systems of Medicine (ISM) in the provision of reproductive and child health services, and in reaching out to households.
- Promote vigorously the small family norm to achieve replacement stages of TFR.
- Bring in relation to the convergence in implementation of related social sector programs so that family welfare becomes a people centered programme.

Strategic themes

- Decentralized Planning and Programme Implementation
- Convergence of Service Delivery at Village Stages
- Empowering Women for Improved Health and Nutrition
- Child Health and Survival
- Special efforts for Under-Served Population Groups who are people living in Urban Slums, Tribal Communities, Hill Area Populations, Displaced and Migrant Populations, Adolescents. Efforts to augment participation of men in Planned Parenthood
- Diverse Health Care Providers including private parishioners, private hospitals, NGOs, etc.
- Collaboration With and Commitments from Non- Government Organisations and the Private Sector
- Mainstreaming Indian Systems of Medicine and Homeopathy
- Contraceptive Technology and Research on Reproductive and Child Health (xi) Providing for the Older Population
- Information, Education, and Communication

Legislation

. It is recommended that the 42nd Constitutional Amendment that freezes till 2001, the number of seats to the Lok Sabha and the Rajya Sabha-based on the 1971 Census be extended up to 2026.

Public support

Demonstration of strong support to the small family norm, as well as personal instance, through political, community, business, professional, and religious leaders, media and film stars, sports personalities, and opinion makers, will enhance its acceptance throughout society. The government will actively enlist their support in concrete ways.

New structures

- National Commission on Population: A National Commission on Population, presided in excess of through the Prime Minister, will have the Chief Ministers of all states and UTs, and the Central Minister in charge of the Department of Family Welfare and other concerned Central Ministries and Departments, for instance Department of Woman and Child Development, Department of Education, Department of Social Justice and Empowerment in the Ministry ofHRD, Ministry of Rural Development, Ministry of Environment and Forest, and others as necessary, and reputed demographers, public health professionals, and NGOs as members. This Commission will oversee and review implementation of policy.

The Commission Secretariat will be provided through the Department of Family Welfare.

- State/ UT Commissions on Population
- Coordination Cell in the Planning Commission
- Technology Mission in the Department of Family Welfare

Funding

Funding to obtained form a variety of sources including international sources, national governments, state governments, NGOs etc. Promotional and motivational events for adoption of the small family norm, the following promotional and motivational events will be undertaken:

- Panchayats and Zila Parishads will be rewarded and honored for exemplary performance in universalizing the small family norm, achieving reductions in infant mortality and birth rates, and promoting literacy with completion of primary schooling.
- The Balika Samridhi Yojana run through the Department of Women and Child Development, to promote survival and care of the girl child, will continue. A cash incentive of Rs. 500 is awarded at the birth of the girl child of birth order 1 or 2.
- Maternity Benefit Scheme run through the Department of Rural Development will continue. A cash incentive of Rs. 500 is awarded to mothers who have their first child after 19 years of age, for birth of the first or second child only. Disbursement of the cash award will in future be connected to compliance with ante-natal check up, institutional delivery through trained birth attendant, registration of birth and BCG immunization.
- A Family Welfare-connected Health Insurance Plan will be recognized. Couples below the poverty row, who undergo sterilization with not more than two living children, would become eligible (beside with children) for health insurance (for hospitalization) not exceeding Rs. 5000, and a personal accident insurance cover for the spouse undergoing sterilization.
- Couples below the poverty row, who marry after the legal age of marriage, register the marriage, have their first child after the mother reaches the age of 21, accept the small family norm, and adopt a terminal method after the birth of the second child, will be rewarded.
- A revolving fund will be set up for income-generating activities through village-stage self help groups, who give community-stage health care services.
- Creches and child care centers will be opened in rural areas and urban slums. This will facilitate and promote participation of women in paid employment.

- A wider, affordable choice of contraceptives will be made accessible at diverse delivery points, with counseling services to enable acceptors to exercise voluntary and informed consent.
- Facilities for safe abortion will be strengthened and expanded.
- Products and services will be made affordable through innovative social marketing schemes.
- Local entrepreneurs at village stages will be provided soft loans and encouraged to run ambulance services to supplement the existing arrangements for referral transportation.
- Increased vocational training schemes for girls, leading to self-employment will be encouraged.
- Strict enforcement of Child Marriage Restraint Act, 1976.
- Strict enforcement of the Pre-Natal Diagnostic Techniques Act, 1994.
- Soft loans to ensure mobility of the ANMs will be increased.
- The 42nd Constitutional Amendment has frozen the number of representatives in the Lok Sabha (on the foundation of population) at 1971 Census stages. The freeze is currently valid until 2001, and has served as an incentive for State Governments to fearlessly pursue the agenda for population stabilization. This freeze needs to be extended until 2026.

Socio-Demographic goals in 2010

- Implementation in totality of the Minimum Needs Programme in scrupulous, universalisation of primary education and reduction in the drop-out rates of primary and secondary school students, both boys and girls, abolition of child labour and priority to primary health.
- Reduction in the incidence of marriage of girls below the age of 18 years to zero.
- Augment in the percentage of deliveries mannered through trained personnel to 100 percent.
- Reduction in maternal mortality rate to less than 100 per 100,000 live births.

Family Welfare Programmes through Five Year Plans

You have read in relation to the Family Planning Policies urbanized through the Government of India throughout the Plan periods. Now let us look at the Family Welfare Programme urbanized through the Five Year Plans. The country is committed to attaining the goals of "Health for All" and a "Net Reproduction Rate of Unity" through the year 2000 A.D. through the universal provision of comprehensive primary health care services to all and an easy access to family welfare planning and maternal and child health facilities. The National Family Planning Programme is the expression of the communal concern for the population problem. After 10 years of its introduction, the Department of Family Planning was organized at the Centre. Now it is described the Department of Family Welfare. At the State or Union Territory stage, there are directorates of Family Welfare Planning. In excess of the years, the programme has evolved a nationwide physical infrastructure and vast reservoir of skills.

Family Planning Programmes under the First Five Year Plan (1951-56).

While formulating the First Plan, it was assumed that the population would continue to grow at the rate of 1.25 percent per annum. So, under the Medical and Public Health Plan, a component entitled Family Planning and Population Control was incorporated for the first time with an allocation of Rs. 6.5 million for this action. The main appeal for family planning was based on thoughts of health and welfare of family. It was understood that family limitation or spacing of children was necessary and desirable in order to secure better care and upbringing of the children. It was then firmly believed that all progress in this field depended on creating a sufficiently strong motivation in favor of family planning in the minds of people. After that it would only remain to give the necessary advice and service, based on acceptable, efficient, harmless, and economic methods. For carrying out the programme of family planning two committees were constituted one to deal with population policy and the second for research and for framing programmes relating to family limitation.

Thus, India was the first country in the world to have an official policy on population and to a lunch a National Programme of Family Planning in 1952. The Family Planning programmes in this plan were expected to obtain. 1) an accurate picture of the factors contributing to the rapid population increase, 2) to discover appropriate techniques of family planning, 3) to devise methods through which knowledge of these techniques could be widely disseminated, 4) to make advice on family planning an integral part of the services of government hospitals and public health agencies.

Family Planning Programmes under the Second Five Year Plan (1956-61)

Throughout the Second Plan the strategy was the same as in the First Five Year Plan that is expansion of family planning services facilities through clinics. The budgetary provision for family planning increased from Rs.6.5 million in the First Plan to Rs. 50 million in second Plans. The sharing of contraceptives was extended through Primary Health Centers, Government Hospitals and Dispensaries, and Maternity Homes run through the State Governments. In both the rural and urban areas, contraceptives were issued free to those with a monthly income below Rs. 100, and at half price to those in the Rupees 100-200 income group. The Central Family Planning Board recommended the inclusion of sterilization operations in the family planning programme in hospitals and institutions where facilities lived. An incentive scheme paying Rs.10/- to a sterilization acceptor as compensation for he loss of wages was first introduced in Tamil Nadu followed through other States.

Research and Training Activities

Research activities were extended to the fields of reproductive physiology, demography, and communication action. Considerable progress was achieved at the contraceptive testing units in Bombay under the guidance of the Indian Council of Medical Research and the All India Institute of Hygiene and Public Health in Calcutta. Demographic research centers were set up in Bombay, Calcutta, Delhi, and Trivandrum. The United Nations Regional Demographic Training and Research Centre at Bombay, recognized in 1956, became a reputed centre for training students in Demography and population studies from dissimilar countries of Asia and the Pacific region. Many valuable field investigations were accepted out throughout 1951-61 such as the India-Harvard Ludhiana Population Study, the Mysore Population study, and the Lodhi Colony Study in Delhi and the Singur Study in Calcutta. A broad-based training programme was urbanized which incorporated establishment of centers for family planning. Family Planning was incorporated in the normal training programme of a number of training institutions for doctors and medical auxiliaries.

In this plan, a Central Board for Family Planning and Population Troubles was set up at the national stage. Its responsibilities were to take care of extension of programme, training of personnel, and organizing biomedical and demographic researches. It was also responsible for carrying out inspection/ supervision of governmental and non-governmental agencies getting grants, monitoring, and evaluation etc.

Family Planning Programmes under the Third Five Year Plan Period (1961-66)

The Third Five Year Plan document sounded a note of concern that the Family Planning programme was a mainly hard one to carry out and it raised troubles of great complexity. It was realized that, sustained and rigorous efforts were described for in excess of a fairly extensive period before family planning could become a popular movement and part of the accepted attitudes of the people usually. Throughout the Third Plan period, the programme was strengthened further and an expenditure of in relation to theRs.250 million was incurred. The basically clinic oriented approach throughout the first two plans was replaced through an extension education through the network of primary health centers and sub centers in the urban areas.

The change in strategy involved utilization of interest and influential local leader in villages for promoting a small-family norm and carrying the message of family planning to the couples. The objectives of this extension approach, which continues to be a pervasive methodology in the Indian Family planning programme to date, are:

- Creation of a group norm of a small family size in every community through educating and involving opinion leaders,
- Providing information to every eligible couple on accessible contraceptive methods, and
- Making provision for contraceptive service facilities in a socially and psychologically acceptable manner. In the clinic approach, the family planning personnel wait for eligible couples to come to their seeking advice and supplies. In the extension approach the crucial task of identifying, informing and motivating the eligible couples for family planning was given to the peripheral health workers, particularly to the auxiliary nurse midwives (anms) and family planning health assistants (fphas). In respect of advocates on methods of family planning, the 'cafeteria approach' was adopted, leaving the choice of the method to an acceptor. The responsibility of sharing of simple contraceptives and giving general advice on family planning was given on a much superior scale to voluntary organizations, paramedical personnel and to extension educators, trained in family planning.

The family planning programme was viewed mainly as a positive policy instrument for achieving the demographic goal. It was not measured as a social welfare measure for improving the status of women in the country or helping couples to space and limit the number of children according to their desire. The programme gained momentum in 1966 when a Department of Family Planning was constituted in the Ministry of Health and Family Planning at the Centre.

It was to provide technical and administrative direction and guidance to the programme and to bring in relation to the effective co-ordination of its several facets. The emphasis was placed on time-bound and target oriented programmes. The Third Five Year Plan stated that "the objective of stabilizing the growth of population in excess of a reasonable period necessity is at the very centre of planned development".

Plan-Holiday (1966-69)

The period 1966-69 was termed as a "plan-holiday" when the earlier programmes were sustained with annual budgeting and target setting. The programme was integrated with the health programme in the country such as integrated with public health programme as maternal and child health (MCH) services operated through the primary health centers (PHCs) in rural areas and urban family welfare planning centers (UFWPCs) in town and cities. The expenditure throughout this period increased to Rs. 704.6 million. Two characteristics that characterized the programme at this stage were:

- The pattern of personnel to be deployed in the programme was decided at the national stage, purely on the foundation of the population size of a phc, district or state
- The choice of the methods accessible to couples was limited with emphasis on IUD and sterilization, although as a matter of policy, all methods were to be made accessible to the couples leaving the choice entirely to them.

Family Planning Programmes in the Fourth Five Year Plan (1969-74)

In the fourth five year plan, family planning was incorporated in the middle of the programmes of the highest priority. A numerical target was set for reducing the crude birth rate from 39 through the end of the Plan period and to 25 through 1979. These demographic goals were translated into targets of family planning acceptors to be recruited under the programme. To achieve these targets, a concrete programme was drawn up for expanding the facilities. It was through providing services and expansion of motivational and educational characteristics through the mass media. An outlay of Rs.3, 300 million was made in this Plan for the programme and the actual expenditure was Rs. 2884.3 million. It is estimated that as a result of the programme 28 million couples were protected through 1973-74 and the births turned absent throughout this plan period were estimated at 12 million.

Throughout this plan period, the programme for popularization of oral pills was also expanded. Surgical equipments were provided in rural and urban family welfare centers for vasectomy operations and a system for free sharing of condoms was introduced.

Through the beginning of this Plan (1969), sterilization became the major task in the government strategy to be met. Increased emphasis was placed on the adoption of the 'camp approach'. In this approach sterilization operations were accepted out in villages at appropriate locations for conducting surgery and the infrastructure facilities were strengthened in the rural areas. The incentive money was also raised.

At the end of this Plan period the mass-camp approach was replaced through the "mini-lap approach" under which in any one camp not more than 25 persons could be operated. This change in excess of was felt necessary since a large number of complaints were received from persons operated upon in large camps. As a consequence, the programme suffered a set back in 1973-74, when the number of sterilization done declined to 0.9 million from 3.1 million in the previous year.

In 1971, Parliament passed a law liberalizing induced abortions under an Act entitled "Medical Termination of Pregnancies Act" which became effective from 1st April 1972, making it possible for pregnant women to have legal abortion under sure specified circumstances.

Family Planning Programme under Fifth Five Year Plan (1974-79)

This plan witnessed a dramatic rise and fall in family planning acceptance in the country. The Fifth Plan document refixed the demographic goals so as to achieve a birth rate of 30 through 1979 and 25 through 1984. Throughout 1974- 78 a sum of Rs.4, 089.8 million was spent on the programme and it received an enormous boost from the government in 1976 with the announcement of the National Population Policy, a comprehensive policy formulated for the first time. The performance in the family planning programme throughout the year 1976-77 was the best to be realized ever, with a total of 8.26 million sterilizations, and all-time record. The programme suffered a serious set back after 1976- 77 and a revised policy on family welfare was announced in April 1977 (1.4.2). The term family planning was changed to "family welfare" to contain maternal and child health programmes as an integral part of the programme.

Family Welfare Programmes under the Sixth Five Year Plan (1980-85)

The sixth Five Year Plan in its objectives stated that one of the major areas of effort which was incorporated pertains to promoting policies for controlling the growth of population through voluntary acceptance of small family norm. The Plan envisaged the extensive term goal of reducing the Net Reproduction Rate to unity through 1995. for the country as a whole and through 2001 in all the states, This was expected to be made possible through reducing the birth rate to 21 and death rate to 9 and raising the proportion of couples protected through family planning to in relation to the60 percent. It is understood that the prime was required to be reactivated through education and persuasion of people, avoiding any form of coercion.

The small family norm was to be built into the social and cultural ethos of the people. A multi-pronged but integrated approach was advocated. It comprised of: 1) education and employment, particularly of women, 2) eradication of poverty, 3) provision of maternal and child care services including immunization, prophylaxis against anemia and nutrition, 4) building up of health care facilities in rural areas with due attention to control of communicable diseases, 5) promotion of preventive health, 6) water supply and sanitation.

Family Welfare services and supplies were sought to be made accessible on an extended scale through the health infrastructure in the country. All existing channels of communication including governmental extension machinery, voluntary organizations, youth organizations, women's organizations, village opinion leaders etc; were to be fully mobilized for promoting the widespread acceptance of family planning methods.

While formulating the targets for family planning methods, adequate attention was given to raising the stage of acceptance of non-terminal/spacing methods in the contraceptive mix. Throughout the sixth plan, an allocation of Rs. 101.00 million was made for family welfare sector. At the end of the Plan, it was envisages that the couple protection rate due to contraception would go up to 36.6 percent. Throughout 1983, the government of India adopted a National Health Policy under which it was sought to be achieved through universalisation of primary health care and reaching Net Reproduction Rate of Unity through 2000 A.D.

Family Welfare Programme under the Seventh Five Year Plan (1985-90)

The Seventh Five-Year Plan document has declared the approval of the extensive-term demographic policy of reaching a net reproduction rate of 1 through the year 2000.A.D. In terms of specific family planning goals for the Plan period, the following targets have been stipulated.

- Effective couple protection rate (CPR) of 42 percent to be realized through 1990.
- Crude Birth Rate (CPR) 29.01 to be realized through 1990.
- Crud Death Rate (CDR) 10.4 to be realized through 1990.
- Infant Mortality Rate (IMR) 90 per 100 live births through 1990
- Immunization of children-universal coverage; and
- Ante-natal care -75 percent of all pregnant women.

In order to reach the above targets, particularly 42 percent couple protection, the Seventh Plan stipulated 31 million sterilizations, 21.25 million IUD (Intra Uterine Devices) insertions, and 14.5 million users of conventional contraceptive through the end of the Plan year 1989-90.

Inter-sect oral co-ordination and cooperation and involvement of voluntary agencies in the programme are contemplated to be implemented in a better measure in the field of health and family welfare. Community participation is being achieved through the utilization of non-government organizations, informal leaders in the community, political leaders, and other social workers.

Special programmes to reduce the infant mortality rate to the stage of 90 per thousand per year through 1990 have been implemented. Special schemes for the reduction of diseases in the middle of children, such as diarrhea, dysentery and respiratory diseases are being implemented.

A Universal Immunization Programme (UIP) providing immunization for children and oral dehydration therapy for treatment of diarrhea diseases was implemented. The UIP sheltered all districts of the country through 1990. The Seventh Plan has provided an outlay of Rs.32560 million for the family welfare sector. Throughout the Seventh Plan, the oral pill sharing programme had been intensified. Besides, a subsidized marketing programme for promoting the oral pill with brand name MALA-D has been launched utilizing the sharing network and services of selected pharmaceutical companies.

Eighth Five Year Plan (1992-97)

- The vital premises of the Family Welfare Programme till now have been) Acceptance of the family welfare is voluntary.
- The Government's role is to create an environment for the people to adopt small family norm. This is done through spreading awareness, information, and education through ensuring easy and convenient availability of family planning aids and services and through giving incentives for adopting family planning.
- The programme, which is a 100% Centrally Sponsored Scheme has integrated family planning and Mother and Child Health (MCH) services and is being implemented through countrywide network of primary health centers and supporting institutions. Inspite of these efforts the results have been far for encouraging.

The following strategies will be adopted for achieving the goals of family welfare throughout the Eighth Plan.

- Convergence of services provided through several social services sectors.
- Decentralized planning and implementation will be another strategy.
- Panchayati Raj institutions like Gram Panchayat and Zila Parishads, etc., will have to play important role in planning, implementing and administering the programme. The role of the Centre will be limited to general policy planning and coordination, providing technological inputs of local self-government.
- The younger couples, who are reproductively mainly active will be the focus of attention, with necessarily a greater emphasis on spacing methods, although the terminal methods would continue to remain the significant means of birth control.
- The targeted reduction in the birth rate will be the foundation of designing, implementing and monitoring the programme against the current method of couple protection rate.
- The outreach and excellence of family welfare services will be improved through ensuring adequate drugs and other essential supplies at the Sub-centre and PHC through suitably raising the funds for this purpose.
- The whole chain of CHC, PHC and Sub-centers will be equipped to deliver general health and MCH services in an integrated manner with a strong referral support and linkage at the District stage.
- Child survival and safe motherhood initiatives will be vigorously pursued. These initiatives will contain (a) strengthening of Universal Immunization Programme, (b) greater emphasis on Diarrhoea Control Programme and effective implementation of art programme, (c) Acute Respiratory Infections Control Programme, (d) Anaemia Management

Programme and not presently Anaemia prophylaxis, (e) Safe Motherhood Programme with high risk pregnancy approach and (f) intensified effort for training of birth attendants.

- Training will not only aim at providing requisite knowledge and social, but also ensure development of such behavioral attributes that will him conducive to a closer interaction with the community. The methodology, the logistics and the content of training programme will be continuously reviewed. Special programmes would be chalked out for imparting pro-service and in-service training in programme management and IEC activities. To meet the training needs, several training institutions will be strengthened or new ones recognized, through providing adequate funds, staff, equipments and mobility.
- The whole package of incentives and awards will be restructured to make it more purposeful.
- There is an urgent need to secure involvement and commitment of practitioners of all systems of medicine in the Population Control Programme.
- The role of voluntary organisation in a mass movement such as population control is critical for generation of momentum and accelerating the pace of progress.
- As an extrapolation of the concept of voluntary organisations, is the role and place of organized corporate sector which covers almost 20 million workers and their families. Effective methods will he evolved to get the organized sector involved in the implementation of family welfare programme.
- Special efforts will he made to involve the community in the Family Planning Programme. The strategy will he to prepare the community to accept the responsibility, the ownership and the control of
- The village/neighborhood tea shops, pan shops, public sharing system shops, pharmacies, cooperatives, etc., will he utilized for community based contraceptive sale and sharing.
- The social marketing of oral pills as well as for market research and educational activities will be done with help for the Corporate Sector possesses special social and sensitivity.
- Information, Education and Communication, which are critical inputs will be further strengthened and expanded.
- A new thrust in the research and development of methods aimed at regulation of fertility in the male, and of vaccines for fertility regulation, both in the male and female, will be given. Fertility regulation practices such as the use of special herbs through the community particularly in the tribal areas, will also be subjected to research. While intensification of bio-medical research is necessary,

research in social and behavioral sciences to explore the human dimensions is vital. Health systems research to optimize operational framework, to improve the efficiency and effectiveness of the service provided and to evolve cost-effective interventions in several areas of family planning operation, will be given high priority.
- A continuous monitoring, review and evaluation is an essential component for the successful implementation of the programme.
- The family planning programme has a multi-sectoral dimension.

The Ninth Five Year Plan (1997-2002)

The reduction in the population growth rate has been recognized as one of the priority objectives throughout the Ninth Plan period. The current high population growth rate is due to:
- The large size of the population in the reproductive age-group (estimated contribution 60%);
- Higher fertility due to unmet need for contraception (estimated contribution 20%); and
- High wanted fertility due to prevailing high IMR (estimated contribution in relation to the20%).

The enabling objectives throughout the Ninth Plan period, so, will be to reduce the population growth rate through The objectives throughout the Ninth Plan will be:
- To meet all the felt-needs for contraception
- To reduce the infant and maternal morbidity and mortality so that there is a reduction in the desired stage of fertility

The strategies throughout the Ninth Plan will be:
- To assess the needs for reproductive and child health at PHC stage and undertake area- specific micro planning
- To give need-based, demand-driven high excellence, integrated reproductive and child health care. a) meeting all the felt-needs for contraception; and b) reducing the infant and maternal morbidity and mortality so that there is a reduction in the desired stage of fertility

The strategies throughout the Ninth Plan will be:
- To assess the needs for reproductive and child health at PHC stage and undertake area- specific micro planning; and
- To give need-based, demand-driven high excellence, integrated reproductive and child health care.

The programmes will be directed towards:
- Bridging the gaps in essential infrastructure and manpower through a flexible approach and improving operational efficiency through investment in social, behavioral and operational research

- Providing additional assistance to poorly performing districts recognized on the foundation of the 1991 census to fill existing gaps in infrastructure and manpower.
- Ensuring uninterrupted supply of essential drugs, vaccines and contraceptives, adequate in quantity and appropriate in excellence.
- Promoting male participation in the Planned Parenthood movement and raising the stage of acceptance of vasectomy.

Efforts will be intensified to enhance the excellence and coverage of family welfare services through:

- Rising participation of general medical practitioners working in voluntary, private, joint sectors and the active cooperation of practitioners of ISM&H;
- Involvement of the Panchayati Raj Institutions for ensuring inter-sectoral coordination and community participation in planning, monitoring and management;
- Involvement of the industries, organized and unorganized sectors, agriculture workers and labour representatives.

The NDC Committee on Population has recommended that there should be:

- Decentralized area specific planning based on the need assessment
- Emphasis on improved access and excellence of services to women and children
- Providing special assistance to poorly performing states/districts to minimize the inter and intrastate differences in performance

Creation of district stage databases on excellence and coverage and impact indicators for monitoring the programme IICPD has advocated similar approach. Concordance flanked by National (NDC Committee) and International (ICPD) efforts has improved funding and accelerated the pace of implementation of the family welfare programme.

- Decentralized area specific planning based on the need assessment
- Emphasis on improved access and excellence of services to women and children
- Providing special assistance to poorly performing states/districts to minimize the inter and intrastate differences in performance
- Creation of district stage databases on excellence and coverage and impact indicators for monitoring the programme.

Tenth Five Year Plan (2002-2007)

The NDC Sub-Committee on Population recommended that there should be a paradigm shift in the Family Welfare Programme and the focus should be on:

- Decentralized area-specific planning based on need assessment.
- Emphasis on improved access and excellence of services to women and children.

- Providing special assistance to poorly performing states/districts to minimize the differences in performance.
- Creation of district-stage databases on excellence, coverage and impact indicators for monitoring the programme.

Department has drawn up the National Population Policy 2000 (NPP 2000), which aims at achieving replacement stage of fertility through 2010. A National Commission on Population was constituted in May 2000, in row with the recommendations of the NPP 2000. Currently some of the major areas of concern contain:

- The huge inter-state differences in fertility and mortality; fertility and mortality rates are high in the mainly populous states, where almost half the country's population lives;
- Gaps in infrastructure, manpower and equipment and mismatch flanked by infrastructure and manpower in primary health centers (phcs)/ community health centers (chcs); lack of referral services;
- Slow decline in mortality throughout the 1990s; the goals set for mortality and fertility in the ninth plan will not be achieved;
- There has been no decline in the maternal mortality ratios in excess of the last three decades, while neonatal and infant mortality rates have plateau throughout the 1990s;
- The routine service coverage has declined, perhaps because of the emphasis on campaign mode operations for individual components of the programme;
- In spite of the emphasis on training to improve skills for the delivery of integrated reproductive and child health (rch) services, the progress in in-service training has been very slow and the anticipated improvement in the content and excellence of care has not taken place;
- Evaluation studies have shown that the coverage under immunization is not universal even in the best performing states while coverage rates are very low in states like bihar; elimination of polio is yet to be achieved;
- The logistics of drug supply has improved in some states but remnants poor in populous states;
- Decentralized district-based planning, monitoring and mid-course correction utilizing the locally generated service data and Civil Registration has not yet been operationalised.

Approach throughout the Tenth Plan

Throughout the Tenth Plan, the paradigm shift, which began in the Ninth Plan, will be fully operationalised. The shift was from:

- Demographic targets to focusing on enabling couples to achieve their reproductive goals,'

- Method specific contraceptive targets to meeting all the unmet needs for contraception to reduce unwanted pregnancies;
- Numerous vertical programmes for family planning and maternal and child health to integrated health care for women and children;
- Centrally defined targets to community need assessment and decentralized area specific micro planning and implementation of program for health care for women and children, to reduce infant mortality and reduce high desired fertility;
- Quantitative coverage to emphasis on excellence and content of care;
- Predominantly women centered programmes to meeting the health care needs of the family with emphasis on involvement of men in planned parenthood;
- Supply driven service delivery to need and demand driven service; improved logistics for ensuring adequate and timely supplies to meet the needs;
- Service provision based on providers' perception to addressing choices and conveniences of the couples.

Three of the 11 monitorable targets for the Tenth Plan and beyond are:

- Reduction in IMR to 45 per 1,000 live births through 2007 and 28 per 1,000 live births through 2012;
- Reduction in maternal mortality ratio to 2 per 1,000 live births through 2007 and 1 per 1,000 live births through 2012; and
- Reduction in decadal growth rate of the population flanked by 2001-2011 to 16.2.

Path Ahead and Goals Set

Reduction in fertility, mortality and population growth rate is major objectives of the Tenth Plan. These will be achieved through meeting all the felt needs for health care of women and children. The focus will be on improving access to services to meet the health care needs of women and children through:

- A decentralized area-specific approach to planning, implementation and monitoring of the performance and effective mid-course corrections;
- Differential strategy to achieve incremental improvement in performance in all states/districts;
- Special efforts to improve access to and utilization of the services in states/districts with high mortality and/or fertility rates;
- Filling the critical gaps, especially in chcs, in existing infrastructure through appropriate reorganization and restructuring of the primary health care infrastructure;

- Ensuring, that post of specialists in chcs do not remain vacant; upgrading skills and redeploying existing manpower to fill other critical gaps;
- Streamlining the functioning of the primary health care system in urban and rural areas; providing good excellence integrated rch services at the primary, secondary and tertiary care stages and improving referral services;
- Providing adequate supply of essential drugs, diagnostics and vaccines; improving the logistics of supply;
- Well coordinated activities for delivery of services through public, private and voluntary sectors to improve coverage;
- Involvement of pris in planning, monitoring and mid-course correction of the programme at the local stage;
- Involvement of industry in the organized and unorganized sectors, agriculture workers and labour representatives in improving access to rch services;
- Effective use of social marketing to improve access to simple in excess of the counter (otc) products such as ort and condoms;
- Effective IEC and motivation programmes; and effective inter-sectoral coordination.

EVALUATION OF FAMILY WELFARE PROGRAMMES

The National Family Planning Programmes (1951) is the expression of the communal concern for the population problem. After a decade, the Department of Family Planning was organized at the Centre. Now there is a close to universal awareness of family welfare planning, its concepts and techniques. In excess of the years, the programme has evolved a nationwide physical infrastructure and a vast reservoir of skills. At the centre, there is a Department of Family Welfare and at the State and Union Territory stage, there is Directorate of Family Welfare Planning.

Although the National Family Planning Programme started in 1951, it has not made an important impact on reduction of fertility. Reduction in birth rate in excess of the years has fallen much short of the planned targets in excess of the successive Five Year Plans. Throughout 1970, the birth rate did not come down from in relation to the39 to 34, but from 1977 onwards it has been stagnating approximately 33 with a slight fall witnessed after 1984. it has been estimated that, the programme has been able to avert in excess of 106 million births, in the country at a total investment of Rs. 4683 crores (almost) upto the end of 1988-89. Thus Rs. 442 has been spent per birth averted in the programme including the cost of infrastructure. Throughout the last 40 years Indian Health and Family Welfare Programme has grown manifold. At present in relation to the16,000 Primary Health Centers (PHCs) and 113,000 sub-centers are functioning in rural areas for providing Health and Family Welfare Services. Through several studies mannered on the Health and Family Welfare Programme, it is found that although it has attained remarkable organizational accomplishment, the programme as a whole has yet to succeed in either curbing population growth or reducing infant mortality to the desired stages

Accessible statistics also show that, performance of the programme is not uniformly poor all in excess of the country. For instance while States like Kerala, Mahrashtra, and Tamil Nadu have performed quite well in terms of both health and FP programme, the States in the Hindibelt, particularly UP, Bihar, Rajasthan and Madhya Pradesh have failed in their effort to implement the programme effectively and augment contraception to the desired stage. If India has to achieve its target of Health for All through 2000 A.D" and Net Reproduction Rate (NRR) equal to one through 2006, through lowering the infant mortality to less than 60 and birth rate to 21, a serious effort should be made to understand the reasons for the ineffective implementation of the programme, so that corrective events could be taken.

FAMILY PLANNING METHODS AND SPACING BETWEEN LIVE BIRTHS

FAMILY PLANNING METHODS

Family Planning Methods or contraceptive methods through definition are preventive methods to help couples to avoid unwanted pregnancies. There are a number of methods which are commonly used through the people. Let us talk about details on those methods one through one.

Condom

Condom is the mainly widely used barrier device through the males approximately the world. In India, it is recognized through its trade name NIRODH, a Sanskrit word meaning prevention. Condom is getting new attention today as an effective simple "spacing" method of contraception, without side effects. In addition to preventing pregnancies, condom protects both men and women form sexually transmitted diseases. There are two kinds of condoms latex and skin. Latex condoms are through far the mainly widely used. The condom is fitted on the erect penis before intercourse. The air necessity is expelled from teat end to make room for the ejaculation. The condom necessity is held cautiously when withdrawing it from the vagina to avoid spilling seminal fluid into the vagina after intercourse. A new condom should be used for each sexual act. Condom prevents the semen from being deposited in the vagina.

The advantages of condom are:

- Are easily accessible,
- Safe and inexpensive,

- Easy to use, do not require medical supervision.
- No side effects,
- Light, compact and disposable, and
- They give protection not only against pregnancy, but also against sexually transmitted diseases (STD).

The disadvantages are:

- It may slip off or tear throughout intercourse due to incorrect use and
- Interferes with sex sensation locally in relation to the some complain while others get used to it, through repeated use.

Although there is much publicity in relation to the use of condoms to avoid pregnancy and getting infected with STDs and HIV/AIDS you necessity keep in mind that the condom does not guarantee hundred per cent safety. There is certainly a risk involved. There are many reported and confirmed cases of condom failure as a preventive method for pregnancy as well as HIV/AIDS infection.

Diaphragm

The diaphragm is a vaginal barrier. It was invented through a German Physician in 1882. It is a shallow cap made of synthetic rubber or plastic material. It has a flexible rim made of spring or metal. It is significant that a woman be fitted with a diaphragm of the proper size. The diaphragm is inserted before sexual intercourse and necessity remain in place for not less than 6 hours after sexual intercourse. A spermidical jelly is always used beside with the diaphragm.

- Advantages: The primary advantage of the diaphragm is the approximately total absence of risks and medical contraindications.
- Disadvantages: Initially a physician or some other trained person will be needed to demonstrate the technique of inserting the diaphragm into the vagina to ensure a proper fit. After delivery, it can be used only after involution of the uterus is completed. Hence it is not very useful in Indian families especially in rural area where medical assistance and privacy hardly exist. Further repeated or frequent pregnancy is a barrier to regular use of diaphragm.

Intra-Uterine Devices (IUDs)

The IUD is devices used for the control of conception through introducing a foreign body into the uterus. There are two vital kinds of IUD: 'non-medicated' and 'medicated'. Both are usually made of polyethylene or other polymoss. In addition the medicated or bioactive IUDs release either metal ions (copper) or hormones (progestogens). The IUDs are of dissimilar generations such as:

- The non-medicated or insert IUDs _ First generation IUDs
- The copper IUDs _ Second generation IUDs

- The hormone releasing IUDs _ Third generation IUDs
- The medicated IUDs or the second and third generation IUDs were urbanized to reduce the incidence of side effects and to augment the contraceptive effectiveness.

Though, they are more expensive and necessity be changed after a sure time to uphold their effectiveness.

The First Generation IUDs

The first generation IUDs comprise the inert or non-medicated devices. They appear in dissimilar shapes and sizes loops, spirals, coils, rings and bows. Lippes Loop, double-S-shaped device, is a very commonly used IUD in India.

The Second Generation IUDs

A new approach was devised in 1970s through adding copper to the IUD. It was found that, metallic copper had a strong anti-fertility effect. The addition of copper made it possible to develop smaller devices which are easier to fit. There are dissimilar kinds of copper IUDs – Copper- 7, Copper-T and Nova-T. The Indian Council of Medical Research in 1979 recommended to the Department of Family Planning the use of Copper-T. According to the recent reports copper devices have become very popular in India.

Advantages:

- Low expulsion rate,
- Lower incidence of side effects,
- Easier to fit even in nulliparous women,
- Better tolerated through nullipara,
- Increased contraceptive effectiveness, and
- Effective as post-coital contraceptives, if inserted within 3 to 5 days of unprotected sexual intercourse.

The Third Generation IUDs

A third generation IUDs is based on another principle that is release of a hormone. The mainly widely used hormonal device is progestseart, which is a T-shaped device filled with progesterone, the natural hormone. The hormone is released slowly in the uterus. Long-term clinical experience with hormone releasing IUD has shown it to be associated with lower menstrual blood loss and fewer days of bleeding than other copper devices. The hormonal devices would be valuable for women in developing countries in who excess blood loss caused through inert devices have shown to result in important anemia. But these devices are too expensive, to be introduced on a wider scale.

Advantages:

- Simplicity, that is no intricate procedures are involved in insertion; no hospitalization is required.

- Insertion takes only a few minutes.
- Once inserted IUD stays in place as extensive as required.
- Inexpensive
- Contraceptive effect is reversible through removal of IUD.
- Virtually free of systematic metabolic side-effects associated with hormonal pills.
- Highest continuation rate.
- There is no need for the continual motivation required to take a pill daily or to use a barrier method uniformly; only a single act of motivation is required.

Hormonal Contraceptives

Hormonal contraceptives when properly used are the mainly effective spacing methods of contraception. They give the best means of ensuring spacing flanked by one child birth and another. Hormonal Contraceptives currently in use may be classified as follows:

- Oral pills: Combined pill; Progestogen – only pill; Post-coital pill; Once-a-month (extensive acting) pill; and Male pill
- Depot (slow release) formulations: Injections; Subcutaneous implants; and Vaginal rings

Oral Pill

The Pill is given orally for 21 consecutive days, beginning on the 5th day of the menstrual cycle, followed through a break of 7 days throughout which period menstruation occurs. When the bleeding occurs this is measured as the first day of the after that cycle.

The bleeding which occurs is not like normal menstruation, but is an episode of uterine bleeding which from an incompletely shaped endometrial caused through the withdrawal of exogenous hormones. So it is described "withdrawal bleeding", rather than menstruation. If bleeding does not occur, the woman is instructed to start the second cycle one week after the proceeding one. Ordinarily the woman menstruates after the second course of pill intake.

The pill should be taken every day at a fixed time, preferably before going to bed at night. The first course should be started strictly on the 5th day of the menstrual period, as any deviation in this respect may not prevent pregnancy. If the user forgets to take a pill, she should take it as soon as she remembers, and that she should take the after that day's pill at the usual time. If taken according to prescription combined pills are 100 percent effective in preventing pregnancy. There is the benefit of pregnancy prevention and risk of abnormal cycle bleeding. The other kinds of pills are to be taken according to the prescription of a medical practitioner.

Adverse Effects

- Cardio vascular effects: Based on some of the studies mannered in dissimilar parts of the world, it is reported that, women who had taken the pill had a 40 percent higher death rate than women who had never taken the pill. Virtually, all the excess mortality was due to cardio vascular causes, that is myocardial infraction.
- Carcinogenesis: Even though there is no clear proof, the WHO multi-centre Case Control Study on the possible association flanked by the use of hormonal contraceptives and neoplasia, indicated a trend towards increased risk of cervical cancer with rising duration of oral contraceptives.
- Metabolic effect: The metabolic effects incorporated the elevation of blood pressure, the alteration in serum lipids with a scrupulous effect on decreasing high density lipoproteins, blood clotting and the ability to vary carbohydrate metabolism with the resultant elevation of blood glucose and plasma insulin. These effects are positively related to the dose of the progestogen component.
- Other adverse effects:
  Liver disorders
  Effect on lactation
  Effect on subsequent fertility
  Ectopic pregnancies and
  Effect on foetal development.
- Common unwanted effects:
  Breast tenderness, fullness and discomfort
  Weight gain
  Headache and migraine
  Bleeding disturbances.

Beneficial Effects

. The single mainly important benefit of the pill is its approximately 100 percent effectiveness in preventing pregnancy. Women taking oral contraceptives should be advised annual medical checkup.

Depot Formulations

The depot formulations are effective, extensive acting oestrogen free for spacing pregnancies, in which a single administration suffices for many month or years. The injectible contraceptives, sub dermal implants, and vaginal rings come in this category.

- Injectable contraceptives: They offer more reliable protection against unwanted pregnancies than the other barrier techniques.

- Subdermal (subcutaneous) implants: The Population Council, New York has urbanized a subdermal implant recognized as 'Norplant' for long-term contraception. The Norplant ® -2, the Silastic capsules or rods are implanted beneath the skin of the forearm or upper arm. Effective contraception is provided for 5 years. The contraceptive effect of Norplant is reversible on removal of capsules. The main disadvantages, though, appear to be irregularities of menstrual bleeding and surgical procedures necessary to insert and remove implants.
- Vaginal Rings: Vaginal rings containing levonorgestrel have been found to be effective. The hormone is slowly absorbed through the vaginal mucosa, permitting mainly of it to bypass the digestive system and liver and allowing a potentially lower dose. The ring is worn in the vagina for 3 weeks of the cycle and removed for the fourth.

Post Conceptional Methods

- Menstrual Regulation: It consists of aspiration of the uterine contents 6-14 days of missed period, but before mainly pregnancy tests can accurately determine whether or not a woman is pregnant. Some regard menstrual regulation as a very early abortion; others view it as a treatment for delayed periods.
- Abortion: Abortion is theoretically defined as termination of pregnancy before the fetus becomes viable (capable of living independently). This has been fixed administratively at 28 weeks. Abortions are usually categorized as spontaneous and induced. Spontaneous abortions occur once in every 15 pregnancies. They may be measured "Nature's method of birth control". Induced abortions, on the other hand, are deliberately induced. They may be legal or illegal. Illegal abortions are hazardous. They are usually the last resort of women determined to end their pregnancies at the risk of their own lives.

Abortion Hazards

Abortions, whether spontaneous or induced, whether in the hands of skilled or unskilled persons are approximately always filled with hazards; resulting in maternal morbidity and mortality.

The early complications of abortion contain shock, septic condition, uterine perforation, cervical injury, thromboembolism, anesthetic and psychiatric complications.

The late complications contain infertility, ectopic gestation, and increased risk of spontaneous abortion and reduced birth weight.

Other Methods of Family Planning

Abstinence

The only method of birth control which is totally effective is complete sexual abstinence. It is sound in theory, in practice it amounts to repression of a natural force and is liable to manifest itself in other directions such as temperamental changes and even nervous breakdown. So, it can hardly be measured a method of contraception to be advocated to the masses.

Coitus Interruptus

This is the oldest method of voluntary fertility control. It involves no cost or appliances. In this method, the male withdraws before ejaculation, and thereby tries to prevent authentication of semen into the vagina. Some couples are able to practice this method successfully, while others find it hard to manage. The chief drawback of this method is that, the precoital secretion of the male may contain sperm, and even a drop of semen is sufficient to cause pregnancy. Further, the slightest mistake in timing the withdrawal may lead to the authentication of a sure amount of semen. The alleged side-effects (e.g. Pelvic congestion, vaginismus, anxiety neurosis) were highly magnified. It is better than by no family planning methods at all. It is admitted to be true that coitus interrupts beside with abstinence and abortion played a major role in reducing birth rates in the urbanized world throughout the 18th and 19th centuries.

Safe Period (Rhythm Method)

This is also recognized as the "Calendar method", first described through Ogino in 1930. The method is based on the fact that ovulation occurs from 12 to 16 days before the onset of menstruation. The days on which conception is likely to occur are calculated as follows:

Regulation of the period and Safe Period Method

If you have irregular period, with a little treatment you can regulate your period. Then you divide your cycle in 3 equal parts. The infertile days are the 1st and the last parts of cycle. The shortest cycle minus 18 days gives the first day of the fertile period. The longest cycle minus 10 days gives the last day of fertile period. For e.g. if a woman's menstrual cycle varies from 26 to 31 days, the fertile period throughout which she should no have intercourse would be from the 8th day to 12th day of the menstrual cycle, counting day one as the first day of the menstrual cycle. The drawbacks of calendar method are:

- Women's menstrual cycles are not always regular. If the cycles are irregular, it is hard to predict the safe period.
- It is only possible for this method to be used through educated and responsible couples with a high degree of motivation and co-operation.

- Compulsory abstinence of sexual intercourse for almost one half of every month may be described “programmed sex”.
- This method is not applicable throughout the postnatal period.
- A high failure rate.

Natural Family Planning Methods

- Basal body temperature (BBT) method,
- Cervical mucus method, and
- Symptothermic method.

Here the woman employs self-recognition of sure physiological signs and symptoms associated with ovulation as an aid to ascertain when the fertile period begins. For avoiding pregnancy, couples abstain from sexual intercourse throughout the fertile stage of the menstrual cycle. They totally desist from by drugs and contraceptive devices. This is the essence of natural family planning.

- Basal Body Temperature Method: The BBT method depends upon the identification of a specific physiological event —— the rise of BBT at the time of ovulation, as a result of an augment in the production of progesterone. The rise of temperature is very small, 0.3 to 0.5 degree C. When no ovulation occurs (e.g. as after menarche, throughout lactation) the body temperature does not rise. The temperature is measured preferably before getting out of bed in the morning. The BBT method is reliable if intercourse is restricted to the postovulatory infertile period, commencing 3 days after ovulatory temperature rise and continuing up to the beginning of menstruation. The major drawback of this method is that, abstinence is necessary for the whole preovulatory period.
- Cervical Mucus Method: This is recognized as ovulation method. This method is based on the observation of changes in the features of cervical mucus. At the time of ovulation, cervical mucus becomes watery clear becoming raw egg white, smooth, slippery and profuse. After ovulation, under the power of progesterone, the mucus thickens and lessens in quantity. It is recommended that, the woman use a tissue paper to wipe the inside of vagina to assess the quantity and features of mucus. This method requires a higher degree of motivation than mainly other methods. The appropriateness of this method in countries like India, especially in the middle of the rural and poor is doubtful.
- Symptothermic Method: This method combines the temperature, cervical mucus and calendar techniques for identifying the fertile period. If the women cannot clearly interpret one sign, she can ‘double check’ her interpretation with another. To sum up, natural family planning demands discipline, and understanding of sexuality. It is not

meant for everybody. The educational component is more significant with this approach than with other methods.

- Breast-feeding: Field and laboratory investigations have confirmed the traditional belief that, lactation prolongs post partum amenorrhea and provides some degree of protection against pregnancy. No more than 5-10 percent of women conceive throughout lactational amenorrhea, and even this risk exists only throughout the month preceding the resumption of menstruation. Though, one menstruation returns, sustained lactation no extensive offers any protection against pregnancy.
- Birth Control Vaccine: Many immunological approaches for men and women are being investigated. The mainly advanced research involves immunization with a vaccine prepared from beta sub-unit of human chronic gonad tropic (HCG) a hormone produced in early pregnancy. Immunization with HCG would block continuation of the pregnancy. Antibodies appeared in relation to the4-6 weeks and reached maximum after in relation to the 5 months and slowly declined reaching zero stages after a period ranging from 6-11 months. The immunity can be boosted through a second injection. Research on birth control vaccines continues and uncertainties are great.

Terminal Methods (Sterilization)

Voluntary sterilization is a well-recognized contraceptive procedure for couples desiring no more children. Sterilization offers several advantages in excess of other contraceptive methods. It is a one-time method. It does not require sustained motivation of the user for its effectiveness. It provides the mainly effective protection against pregnancy. The risk of complications is small if the procedure is performed according to accepted medical standards. It is cost-effective.

- Male Sterilization: Male sterilization or vasectomy being a comparatively simple operation can be performed even in primary health centers through trained doctors under local anesthesia. In vasectomy, it is customary to remove a piece of vas at least 1 cm after clamping. The ends are ligated and then folded back on themselves and sutured into position so that, the cut ends, face absent from each other. This will reduce the risk of decimalization at a later date. It is significant to stress the person is not immediately sterile after the operation, usually until intermediate period, another method of contraception necessity be used. If properly used vasectomies are 100 percent effective. Vasectomy is a simpler faster and less expensive operation than tubectomy.

- Female Sterilization: Female sterilization can be done as an interval procedure; post partum or at the time of abortion. There are two such commonly used methods.
- Laparoscopy: This is a technique of female sterilization through abdominal approach with a specialized instrument described "laparoscope". The abdomen is inflated with gas (carbon dioxide, nitrous oxide or air) and the instrument is introduced into the abdominal cavity to visualize the tubes. Once the tubes are accessible, the Falope rings (or clips) are applied to obstruct the tubes. This operation should be undertaken only in those centers where specialist obstetrician gynecologists are accessible. The short operating time, shorter stay in hospital and a small scar are some of the attractive characteristics of this operation.
- Minilap operation: Minilaparotomy is a modification of abdominal tubectomy. It is a modification of abdominal tubectomy. It is a much simpler procedure requiring a smaller abdominal incision of only 2.5 to 3 cm mannered under local anaesthesia. It is found to be an appropriate procedure at the primary health centre. It has the advantage in excess of other methods with regard to safety, efficiency, and ease in dealing with complications. Minilap is appropriate for postpartum tubal sterilization.

SPACING FLANKED BY LIVE BIRTHS

Till now you have learnt the dissimilar methods of family planning. Now let us see what is spacing and how does it affect the health of mother and child. Spacing is the interval flanked by two lives births. There are a number of social and cultural factors which have tended to augment the spacing flanked by two births.

The segregation of women after delivery, the taboo of sex relations when the child is young and abstinence on sure religious days are some of them. Prolonged lactation which has been observed may also be a significant factor. The Hindu joint family has also helped in minimizing the frequency of sexual relationship. The prevalence of 'purdah' system did not allow the couples to meet very regularly. The so described rigid rules of social behaviour prevalent in traditional families did not allow undue sexual freedom to the couples. Social customs of the wife making frequent visits to her parental home, having the first child in her mother's home and such other customs have helped to augment spacing flanked by two births.

The above mentioned factors which helped to augment spacing are slowly changing. With increased education, urbanization and economic prosperity, social atmosphere is changing. Widow remarriage is becoming more common; moral restraint is not much observed; and the joint family system is giving way to nuclear families. All these factors affect spacing. The birth of children is a voluntary decision of the couples rather than a culturally oriented phenomenon.

RELIGIOUS VIEWS AND SPIRITUAL GUIDANCE

Our lives are profoundly influenced through advancement in the field of science and technology. As a result of these advancements some people tend to consider religion and belief in God outdated and irrelevant. For them the teachings and heritage of the great religions make little or no sense. It is true to say that, the teachings and traditions of the significant religions have been the conscience keepers of the world. The ethical views of dissimilar religions have always condemned the violation of natural law. All religious accept regulating family size through way of self-control and regulating sexual union flanked by couples. The same way all religions condemn abortion and consider it as murder.

According to Hindu Vedas, abortion is measured to be a more serious sin than the killing of a Brahman. The 'Charka Samhita; a classical work on Hindu Medicine states that conception takes place in the womb through the union of semen and ovum, when the soul, beside with the mind, enters the zygote. The embryo is of unique constitution, because it is a composite of the vital information it receives from both parents. The humanization of the individual takes place at the moment of conception and all future growth is only the actualization of conceptual potency. Based of 'Shastras' and on the principle that the genetic components are complete at conception, the modern Hindu belief is that life begins with conception. Though Hindu ethics condemn abortions usually, it accepts abortion on the grounds of rape, incest and when the mother is at the risk of grave injury or death. This is because, Hindu ethics place greater weight on the maternal rights rather than on the unborn child's right.

The vital teaching of Islam is that, life is a gift of god. Hence the Koran warns men not to interfere with the work of God. It is on this faith that, Muslims usually oppose abortion. The laws of Islam prohibit abortion since the foetus is measured a living being. But as per the doctors, foetus is only a human being after the fourth month. Hence abortion is allowed in general throughout the first ninety days of pregnancy and it is prohibited immediately afterwards. Though, like Hindu ethics, Islamic ethics also permits abortion on the foundation of pregnancy which endangers the mother's life, and where it is the result of a rape that does not result in marriage. The Christian religion also condemns abortion. Christianity encourages couples to have self-control and promote natural family planning methods. Christianity considers abortion a grave sin and calls it murder. At the time of conception a new life has started and abortion is the killing of that new life. The Holy Bible warns emphatically against abortion at any stage of conception.

There are sure myths concerning religious faith, population, growth and family planning. Hindu religion promulgates monogamy and there is no law prohibiting family planning. But Muslim religion allows polygamy and prohibits family planning and hence there is rate of augmentin population in the middle of them.

These myths are to be examined in the light of studies mannered in India. When these above mentioned myths are examined based on scientific study, it is understood that they remain as myths even now. In the middle of Muslims it was 23% in 1980 and increased to 34% (+11%) in 1989. Likewise, in the middle of scheduled castes it was 28% (1980) increased to 39% in 1989. In the middle of Scheduled Tribes it was 33 % in 1980 and sustained the same pattern in 1989 also. Muslims are politically and culturally against the attitude of western concept of sexuality and abortion, rather than on family planning. Christians are very conservative in relation to the artificial family planning methods. But it is deadly against abortion.

Even though, all these religious are not positively promoting family planning, in the State of Kerala where literacy rate is very high, education of women is high and a high index of social development exist, the small family norms has been accepted and has become a way of life in the middle of people of dissimilar walks of life, no matter to what religion they belong.

PREFERENCE FOR MALE CHILD

In India sons are significant from a religious point of view. According to Hindu religion, a man, attains salvation only when a son performs sure rites at his funeral. Preference for a son is not only a religious point of view, sons are measured as providers of security in old age and throughout prolonged illness. There are several other reasons as well. Once married, the girls are measured to belong to their husband's family, and so, they cannot be relied on for support in old age. In fact, there are strong taboos on taking any kind of help from a married daughter. According to Blaikie the reasons for the importance of sons in Indian culture are:

- Sons are required to perform the last funeral rites (sraddha) for their parents. It is motivating to note that in Sanskrit 'put' means hell and 'putra' means literally 'one that saves from hell'.
- Sons, upon marriage, draw dowries for the parents.
- Sons, give economic and emotional security in old age. It is the son, not the daughter who remnants at the parent's home after marriage.
- Sons give income and help in the home and in the fields from an early age.
- Sons bring prestige and local political power (and even protection against the threat of physical force in confrontational situations) to the household, the kindship group and caste.

The question of son survivorship is so, so vital that people do not feel satisfied with presently one son. They seek safety in numbers. Women too have compelling reasons for desiring children, preferably sons. The mainly intense hope of a young woman is that she proves her worth to her husband's family through producing a healthy male child. The birth of a son entitles a woman to respect and status. She and her baby have been conditioned to see their success and destiny in terms of procreation especially of a son. In actual practice it is found that, the above arguments for male child are myth. Now as a result of the disintegration of joint family system and augmentin the number of nuclear families, the expectation on the son to give security at old age is coming down. With regard to the amount of work done through males and females, it is approximately equal. Hence preference for male child is a myth.

MEDICAL TERMINATION OF PREGNANCY AND ISSUES ASSOCIATED WITH IT

ABORTION – DEFINITION AND KINDS

Abortion is made up of two Latin words —— ab- meaning off or absent oriri meaning to be born; abortion means taking absent a human life which would in the normal course of events be born. Abortion has two meanings —— medically it can describe a case of miscarriage, without any outside intervention, occurring within the first three months of pregnancy. Abortions legalized through the Act on the other hand, are those deliberately procured with the intention of terminating the pregnancy, killing the unborn child. The terms abortion and miscarriage are sometimes used a synonyms. They are also used to describe the same happening at an earlier and at later stage of pregnancy. Abortion is restricted so as to describe the case occurring in the first three months of pregnancy and miscarriage to describe one throughout pregnancy from the beginning of the fourth month, until the foetus becomes viable.

Kinds of Abortion

- Spontaneous Abortion (Miscarriage) occurring naturally without any deliberate effort on the part of pregnant women or other persons.
- Indirect Abortion: This is an abortion which occurs as a side effect of treatment given to the mother for some diseases or haemorrhage which is endangering her life. The purpose is not to take absent the child's life. It occurs as the result of the therapeutic events taken to save the mother.
- Artificial or Induced Abortion is taking absent of life for the main purpose of removing the foetus or child. Induced abortion is the result of deliberate effort on the part of the pregnant women or others with the intention of terminating pregnancy.

Methods of Procuring an Abortion

There are several ways of aborting an unwanted foetus. The method chosen depends mainly on the duration of pregnancy to be interrupted as in the first or second trimester (3 months).

First Trimester

In the first trimester (upto 12 weeks) of pregnancy, two broad methods are used.

Medical:

- One single dose of Misetristone pills administered within 40 days of menstrual period, (within 10 days after missed period).

Surgical:

- Menstrual Regulation by MR Syringe (upto six weeks of pregnancy)
- Suction: The mainly regularly used method of abortion is suction, also described Vacuum aspiration. It is used when the women is less than 3 months pregnant. This technique involves sucking out the contents of the uterus through a tube inserted into the uterus through the cervix. This operation can be performed while the women are awake, in five to ten minutes, with little blood loss and a low risk of complications.
- Tent Evacuation with Laminaria Tent (slow dilatation is achieved and evacuation is done later).
- Dilatation and Curettage (D&C): This procedure involves rising the size of the cervical canal through inserting a series of slowly – widening metal dilators. When the opening is wide enough the physician uses a curette (small metal surgical instrument) to scrape the embryo and placenta from the walls of the uterus. Since the D and C operation takes longer and is more complicated than the suction method, the women is put to sleep.

Second Trimester

In the second trimester of pregnancy, only very few methods are usually adopted.

- Intra Uterine Saline Instillation: After twelve weeks, the foetus is large enough to be removed safely through the suction or D&C methods. Pregnancies after this period are removed through inserting extensive needle abdominal and uterine walls into the cavity of the uterus. A concentrated salt solution is inserted into the amniotic sac destroyed, the uterus will contract until the foetus is pushed out into the vagina. Because saline abortion is a major surgical procedure, earlier termination of pregnancy through the suction method or D&C is being recommended through the physicians.

- Hysterectomy: This is used in the later stages of pregnancy when the foetus is too large. The mother is given a general anaesthetic and the uterus is cut open and the foetus is removed. This is usually performed in cases where foetal abnormalities are detected and when all the other methods have failed.

Two other methods of abortion are intra cervical prostaglandin E2, administration and oxytocin induction.

LIBERALIZATION OF ABORTION LEADING TO LEGALIZATION

Prior to 1950s in several countries, laws governing abortions were very restrictive; i.e. abortions were totally prohibited or permitted strictly on medical grounds. But laws could not prevent illegal abortions which were done mostly through non-medical persons and in unhygienic circumstances.

Consequences of Illegal Abortion

Illegal abortions regularly led to complications such as perforations of the uterus, hemorrhage and infection requiring gynecological care and hospitalization. In countries such as Venezuela, Nigeria, Chile and a few other countries sepsis due to illegal abortion was the leading cause of maternal deaths. Illegal and partial abortion was a major strain on medical possessions. It is also reported that in countries with restrictive abortion laws, the poor were made to suffer more than the affluent, the better educated and the urban elite. The poor, with no possessions to obtain safe but expensive abortions and with little contacts or knowledge in relation to the places where such services were accessible, were often driven into the hands of unqualified abortionists which resulted in severe complications and deaths. Such loss of lives, impairment to mother's health and wastage of medical infrastructure as well as compassion for the poor led to the demand in many countries for liberal abortion laws.

Attempts to Legalize Abortion

The Soviet Union was the first country to legalize abortion. In 1920, Lenin's government enabled women in first trimester pregnancy (less than 12 weeks of pregnancy) to obtain abortion on request. Thereafter several countries introduced liberal clauses in their abortion laws from 1930 onwards. Presently, the laws on abortion vary from abortion on request at one extreme to total prohibition on the other. The four main countries which have liberalized abortion laws for medical termination of pregnancy are China, India, USA and Soviet Union.

Medical Termination of Pregnancy Act, 1971 (MTP)

In India, prior to 1972, abortion was illegal except to save the life of pregnant women. In 1964, a committee under the Chairmanship of Shantilal Shah was constituted to study the question of liberalizing abortion laws. Based on the report submitted through this committee in 1966, the Medical Termination of Pregnancy (MTP) Act was passed through the Parliament in 1971 and came in to force all in excess of the country in April 1, 1972 (except in Jammu & Kashmir where it came into effect from November 1, 1976).

This act is one of the mainly liberal in the world and has replaced one of the mainly rigid laws sections 312 of Penal Code, 1860 and the Code of Criminal Procedure 1898. The Medical Termination of Pregnancy Act 1971 lays down three norms for terminating a pregnancy.

The circumstances under which a pregnancy can be terminated under the MTP Act 1971:

There are 5 circumstances that have been recognized in the Act.

- Medical – where continuation of the pregnancy might endanger the mother's life or cause grave injury to her physical or mental health.
- Eugenic – where there is substantial risk of the child being born with serious handicaps due to physical or mental abnormalities.
- Humanitarian – where pregnancy is the result of rape.
- Socio-economic—— where actual or reasonably foreseeable environments (whether social or economic) could lead to risk of injury to the health of the mother.
- Failure of contraceptive devices – the anguish caused through an unwanted pregnancy resulting from a failure of any contraceptive device or method can be presumed to constitute a grave mental injury to the mental health of the mother. This condition is a unique characteristic of the Indian Law virtually allows abortion on request, in view of the difficulty of providing that a pregnancy was not caused through failure of contraception.

The written consent of the guardian is necessary before performing abortion in women under 18 years of age, and in lunatics even if they are older than 18 years.

The person or persons who can perform abortion

The Act provides safeguards to the mother through authorizing only a registered medical practitioner having experience in gynecology and obstetrics to perform an abortion where the length of pregnancy does not exceed 12 weeks. Though, where the pregnancy exceeds 12 weeks and is not more than 20 weeks the opinion of two registered medical practitioners is necessary to terminate the pregnancy.

Where abortion can be done

The Act stipulates that no termination of pregnancy shall be made at any place other than a hospital recognized or maintained through the Government or a place approved for the purpose of this Act through Government. Abortion services are provided in hospitals in strict confidence. The name of the abortion seeker is kept confidential, since abortion has been treated as a statutory personal matter.

MTP Rules (1975)

Rules and Regulations framed initially were altered in October 1975 to eliminate time-consuming procedures involved in MTP and to make services more readily accessible. These changes have occurred in three administrative areas.

Approval through Board

Under the new rules, the Chief Medical Officer of the District is empowered to certify that a doctor has the necessary training in gynecology and obstetrics to do abortions. The procedure of doctors applying to Certification Boards was removed.

Qualification required doing Abortion

The new rules allow for registered medical practitioners to qualify through on the spot training. The doctor may also qualify to do MTPs under the new rules if he/she has one or more of the following qualifications which are similar to the old rules:

- 6 months housemanship in obstetrics and gynecology.
- a post-graduate qualification in OBG
- 3 years of practice in OBG for those doctors registered before the 1971 MTP Act was passed.
- 1 year of practice in OBG for those doctors registered on or after the date of commencement of the Act.

The Place where Abortion is Performed

Under the new rules, non-governmental institutions may also take up abortions provided they obtain a license from the Chief Medical Officer of the district, thus eliminating the requirement of private clinics obtaining a Board License.

Limitations of the Act

- The provisions for abortion under the Act such as pregnancy caused through rape, failure of a contraceptive etc. need not necessarily constitute grave injury to her mental health.
- Extending the period of pregnancy from 12 weeks to 20 weeks of pregnancy in sure situations gives more scope for abortion takers.
- The value of human life is challenged
- No consideration is given to the right of the foetus.

The New Act is indeed a liberal piece of legislation compared to the Old Act of 1860. The Act has been mainly adopted to eradicate a large number of criminal and clandestine abortions, which caused considerably mortality and morbidity in the middle of pregnant women. Though, the utility of the new act will depend to large extent on the number of localities that are provided with clinical facilities for abortion especially in rural areas, knowledge in relation to the facilities in the middle of the people and also on the attitude of the physicians. Due to inadequate data on the socio-economic status of women who had legal abortions, it is not clear whether liberalization has really helped the poor sections of the society.

ISSUES AND CONTROVERSIES ASSOCIATED WITH LEGAL ABORTIONS

Abortion has been a hotly debated issue. Both pro (for) and anti-abortion (against) groups have lobbied intensively on the local, state, national, and international stages. Abortion raises some hard questions to which there are no simple answers. Abortion issues may be divided into 1) Physical and Medical, 2) Psychological, 3) Social, and 4) Moral and Religious

Physical and Medical Issue

A woman is made physically and psychologically for motherhood. This is the vital fact of her life. If this procedure of becoming a mother is suddenly stopped, the shock will have its effect. This effect may be physical or mental immediate or extensive-term. If a pregnancy is to be terminated, it should be terminated as soon as possible. This is of special necessity from a biological and medical point of view. Medical complications arise due to patients spending so little time under observation after their initial operation and so troubles of infection tend to set in. A woman who has undergone an abortion is also more likely to have subsequent children both physically and mentally handicapped. Damage to the wall of the uterus can affect the normal development of the placenta through which the baby takes its nourishment. There is an extensive-term medical effect of aborting first pregnancy. Repeated abortions may be associated with a later inability to conceive or to carry a child to full term and with several birth complications.

Abortion should only be a backup measure not the primary method of birth control and they urge fuller usage of contraceptives in the middle of all sexually active person so that repeated abortions are not necessary.

Emotional Issue

There is emotional and physical unrest experienced throughout the first few weeks of pregnancy. Mainly women even experience feelings of rejection of the pregnancy at this time. It is at this time that the expectant mother may be subjected to maximum pressure to agree to an abortion. Each individual is dissimilar. For some, abortion provides great relief with little or not disturbance. For the others, the experience is upsetting. The key factor seems to be whether the woman wants an abortion or whether she is hesitant. Being refused an abortion and forced to bear an unwanted child can lead to psychiatric symptoms. But the woman who has health troubles and has to have an abortion or who is persuaded to have an abortion against her better judgment is also more likely to show negative psychological reactions following the operation.

The common psychological reactions can be minimized if the decision is solely that of the women. So, Abortion Counseling can assist her in making the best decision she can live with and working her feelings ahead of time.

Social and Realistic Issues

Legalized abortion saves lives through reducing the number of illegal attempts. Antiabortionists emphasize their fears that without any restriction, except the individual woman and her conscience, an 'Abortion Mentality' develops so that abortion becomes too common and are performed too easily or for reasons that are not serious: For instance teenage pregnancy has become a common occurrence in the middle of college students in cities with a free access to abortion facilities. Even unplanned pregnancies in normal families, for want of spacing flanked by children, especially in the middle of the career women, are also some of the common groups taking up abortions. Thus, the majority of abortions today are not for medical cause, but for personal, social and economic reasons that have to do with woman's life situations and not with her health.

Moral Issues (Rights of the Unborn Child)

Much of the controversy in relation to the abortion has centered approximately the moral issues involved. In ordinary justice, the child has as much claim as the mother to life and should have even more claim to legal protection of its right, since it is incapable of defending itself.

Rights of the Unborn Child

The UN declaration on the Rights of the child maintains that "the Child through cause of his physical and mental immaturity, needs special safeguards and care including appropriate legal protection before as well as after birth". They emphasize the right to life of the foetus and that no individual or state should deprive the foetus of its constitutional and moral rights to live. The Anti abortionists claim that science has proven beyond any reasonable doubt that human life begins at fertilization. The foetus from the beginning has its own life, is a totally new human being, a new person, with a genetic code quite separate from the genetic code of its parents. The only life generated through human beings is human life. That new life is totally there at fertilization, lacking only development and growth. Abortion always takes absent an innocent's already existing life.

Human life is the highest form of life on earth. The after that generation depends on the existing society for its survival. On the other hand, the pro-abortion lobby emphasizes that the moral and legal rights of other parties necessity also be measured, not presently those of the foetus.

What in relation to the rights of the mother, father, other family members? Should these lives be sacrificed for the sake of the child? Is it right to let the foetus live but to let the mother die so that her husband and other children are deprived of her love?

The constitution guarantees equal protection under the law. Is it moral to force a woman to bear a baby she doesn't want, can't care for, or that might be deformed? Is it moral to insist that an unwanted child be born into the world and then to suffer all of its life because it was never wanted. Who has the right to decide? It is obvious to you know that the moral dilemmas raised through the abortion issue are not easy to solve.

RELIGIOUS VIEWS

The whole gamut of laws related to life is of several kinds ——divine law, natural law, secular or civil law, religious law and the like. The mainly significant law related to life is the Divine Law or the Law of God. God is the giver and the author of life.

No human being on earth has the power to destroy life. Abortion is equal to murder of the human being, a person created in the image and likeness of God and so a grave sin against God. The ethical views of dissimilar religions have always condemned the violation of natural law, the laws of the creator. Keeping this in view, let us look at the teachings of at least three major religions on human life and related issues, specifically on abortion which has posed the mainly serious ethical troubles to the modern world.

The Hindu View

According to Hindu religion, a woman who undergoes an abortion in this, life becomes barren in her subsequent lives. Hindu scriptures condemn abortion and consider it as murder. According to Hindu Vedas, abortion is measured to be a more serious sin that killing of a Brahmin. According to Shastras, life begins after the fourth month in the womb, when a ceremony is performed, blessing the foetus which is already living. Though Hindu ethics condemn abortion usually, it accepts abortion on the grounds of rape, incest, and when the mother runs the risk of grave injury or death. But still some of the Hindu religious leaders oppose such abortion on the ground that it is an act of interference with the Karmic development of the child.

The Islamic View

The Holy Quran warns men not to interfere with the work of God. It is one of the vital teachings of Islam that life is a gift of God and as such, no man has any right to commit any kind of act that is detrimental to, and extinguishing life. It is on this faith that Muslims usually oppose abortion, as it is felt that it amounts to extinguishing of life. The laws of Islam under sure circumstances prohibit abortion when the foetus is animated and measured a living being. As per the doctors of the law, the foetus is only a human being after the fourth month. Abortion is allowed in general throughout the first ninety days of pregnancy, and it is prohibited immediately afterwards. Though, like Hindu ethics, Islamic ethics also permit abortion on the foundation of pregnancy which endangers the mother's life and where it is the result of a rape that does not result in marriage.

Christian Proposition

The Christian church in the first centuries after Christ, forbade abortion under all circumstances from the moment of conception and abortion at any time is measured a grave sin. Direct abortion of a foetus before viability is never lawful because this is to kill an innocent human being, who has a right to live. Every human being, even a child in its mother's womb has a right to life directly from God, and not from Parents or from any human society or power. Hence there is no human power, no medical eugenic, social, economic or moral indication that can offer or produce a judicial title to the deliberate disposal of an innocent human life. There is an argument that the soul does not exist until the foetus really resembles a human being 'in form'. But from the time when the male sperm unites with the ovum fertilizing it, (conception) the minute embryo contains all the features which will make it a unique person. Hence the Christians believe that the soul enters at the moment of conception so that at whatever stage of growth, the new life is human.

View in Favor of and against Abortion

We have so far discussed the issues associated with the legal termination of pregnancy. The views of the two groups: Pro abortionists and anti abortionists are strong in their own ways and the debate seems to be unresolvable. It is likely to persist for sometime generating high sentiments on both sides. We will furnish here very briefly the views favoring legal abortion.

- Women should have the right to control their own bodies.
- No child should be brought into the world unwanted.
- Legal abortion should be mannered in authorized medical settings in which considerable care is taken to avoid harming the mother physically or psychologically.
- Women necessity has the option of a safe, legal abortion if they desire.

View against Legal Abortion

- Foetus is a living being and so its right to life necessity is respected – no one has the moral right to take that life.
- Persons other than the mother have rights as far as the unborn child is concerned – child itself and the father.
- Because the foetus is unable to defend itself, opponents of abortion believe that others are obligated to defend the foetus against the efforts of those who want to "kill it".

The mainly significant opposition to abortion is from organized religious groups who address abortion issue as that of questioning the ultimate power of God the Almighty.

Viable Alternatives to Unplanned Pregnancies

- Family Life Education should help the couples for a planned parenthood.
- Make people realize the value of human life to protect it through instilling in them the love of God.
- Girls in moral danger should be protected.
- Effective use of contraceptives should be promoted.
- Teenagers should be helped to imbibe the spiritual values in them and be aware of the consequences of ending countless innocent human lives through premarital sexual experiences.

Bibliography

Aashima Arora and Hemant Kapoor: *Home Management*, Bio-Green Books, Delhi, 2012.

Afzal Sharieff, Masood Ali Khan and A. Balakishan: *Land Policy and Urban Housing*, Sarup Publication, Delhi, 2010.

Amrish Kumar Saxena: *Sociological Dimensions of Urban Housing and Development*, Commonwealth Publication, Delhi, 2004.

Bhaskar Majumder: *Housing on the Hills in India*, Concept Publication, Delhi, 2011.

Bhaskar Majumder: *Rural Housing : Policies and Practices*, Rawat Publication, Delhi, 2007.

Braja Bandhu Padhiari and Prakash Chandra Misra: *Rural Housing Finance*, Discovery Publishing House, Delhi, 2004.

Brian C. Aldrich and Ranvinder S. Sandhu: *Housing for the Urban Poor in Developing Countries*, Rawat Publication, Delhi, 2015.

C. Gangaiah and G. Rajesh Kumar: *Rural Housing Schemes and Policies : A Study*, Serials Publications, Delhi, 2012.

DE CHIARA: *Time-Saver Standards for Housing and Residential Development*, Tata McGraw-Hill, Delhi, 2009.

Girish K. Misra and P.S.N. Rao: *Housing Legislation in India : Policies and Performance*, Kanishka Publication, Delhi, 2000.

L. Rajani and P. Mohan Reddy: *Housing Finance by Commercial Banks*, Serials Publications, Delhi, 2011.

M. Lakshmi Narasaiah and B. Uma Jyothi: *Housing Programmes for the Poor*, Sonali Publication, Delhi, 2009.

Maneesha Shukul, Veena Gandotra and Sarjoo Patel: *A Compendium of Research in Home Management*, Authorspress, Delhi, 2011.

Peer Smets: *Housing Finance and the Urban Poor*, Rawat Publication, Delhi, 2004.

Prabhu Dayal Choudhary: *Rural Housing and Infrastructural Problems in India*, Prateeksha Publication, Delhi, 2010.

Priya Bhargav and Tara Chand: *Home Management and House Keeping*, Commonwealth Publication, Delhi, 2005.

Pushpa Chakravorty: *Home Management*, Pointer Publication, Delhi, 2007.

Rishi Muni Dwivedi: *Urban Development and Housing in India 1947 to 2007*, New Century Publication, Delhi, 2007.

ROBERT MATTISON: *Web Warehousing and Knowledge Management*, Tata McGraw-Hill, Delhi, 2007.

Ruchi Jain: *Housing Finance in India : Turning Dreams into Reality*, A.K. Publication, Delhi, 2011.

S. Giriappa: *Housing Finance and Development in India*, Mohit Publication, Delhi, 1998.

Sawalia Bihari Verma; Sant Gyaneswar Prasad Singh and Shib Kumari Singh: *Rural Infrastructure : Sanitation, Housing, Healthcare*, Sarup and Sons Publication, Delhi, 2008.

Shilaja N. Ambedkar: *Rural Housing : Agro Socio Economic Impact*, Agrobios Publication, Delhi, 2000.

Suba Singh: *Housing Geography*, ABD Publishers, Delhi, 2011.

V K Kohli: *Housing Finance Agencies in India*, Deep and Deep Publication, Delhi, 2007.

Veena Gandotra, Maneesha Shukul, Neena Jaju and Neerja Jaiswal: *Housing : Changing Needs and New Directions*, Authors Press, Delhi, 2009.

Index